22 Advances in Biochemical Engineering

Managing Editor: A. Fiechter

Space and Terrestrial Biotechnology

With Contributions by
A. Cogoli, I. I. Inculet, P. Martin,
K. Schügerl, A. Tschopp, J. E. Zajic

With 136 Figures and 29 Tables

Springer-Verlag
Berlin Heidelberg GmbH 1982

ISBN 978-3-662-15356-7   ISBN 978-3-540-39084-8 (eBook)
DOI 10.1007/978-3-540-39084-8

© by Springer-Verlag Berlin Heidelberg 1982
Originally published by Springer-Verlag Berlin Heidelberg New York in 1982
Softcover reprint of the hardcover 1st edition 1982

Library of Congress Catalog Card Number 72-152360

2152/3020-543210

Table of Contents

Biotechnology in Space Laboratories

Augusto Cogoli, Alex Tschopp
Laboratorium für Biochemie, ETH-Zentrum, CH-8092 Zürich, Switzerland

The advent of the Space Shuttle and of the Spacelab will open new perspectives to biotechnology in space. The objectives of this review are: a) to present an overview on the technological and scientific aspects of biological experiments performed on the past US and Soviet space missions, b) to describe the facilities offered by Spacelab in the future, c) to give practical information on the requirements of flight hardware and on the limits in weight, energy and crew-time. Experiments on Apollo and on the Apollo-Soyuz Test Project have shown that the weightless environment offers advantageous conditions to the processing of mammalian cells. Soviet investigations demonstrated that artificial gravity attenuates some of the disturbing effects of spaceflight observed on animals and plants. An extensive program of preliminary studies should precede large-scale biotechnological applications: Suitable hardware has to be developed in collaboration with ESA and NASA, biological objects should be selected as potential candidates for bioprocessing in space. The preparation of investigations in space should be accompanied by ground high-g simulations in centrifuges and by compensating gravity in clinostats. The exploitation of space resources and the establishment of space colonies is becoming a realistic goal for the next decades.

1 Introduction

Over twenty years of manned spaceflights have shown that man can survive and work in space for a prolonged period of time (more than 6 months). The main objectives of the past missions were to show that man can adapt himself to the new environment and to explore the solar system. Astronomy and astrophysics were the disciplines which most profited from the scientific space program. In comparison, facilities for life sciences were less sophisticated and were flown on a limited number of missions.

With the advent of the Space Shuttle and of the Spacelab we are at the beginning of a new era in space technology: Time has come for the exploitation of space resources.

It is known that weightlessness provides favorable conditions for the production of high-quality materials like metal alloys and glasses. Similarly, preliminary experiments indicate that purification of specialized human and animal cells is sometimes easier to perform in space than in ground laboratories. Therefore, material sciences and bioprocessing are becoming new important disciplines in space sciences. An extensive program of research in space biology will start as soon as Spacelab becomes operational. A basic study on the survival and adaptation of living systems to the space environment has to be performed in parallel with applied research. Every biological experiment in space is a technological challenge since flight hardware must comply with the strict safety and engineering specifications set by the space agencies on the one hand and with the biological requirements of living systems on the other.

The objective of this review are:
— To give a summary of the results and describe the equipment of experiments in space biology performed on past missions.
— To discuss the technological aspects of the preparation and execution of life science projects in space. In particular, we describe our own experiments to be performed on Spacelab;
— to draw the attention of the reader to data and observations often not easily available;
— to stimulate the scientific community to participate in the activities in space taking profit of the facilities offered by the Spacelab and other space stations;
— to analyze the potential benefits of space research;
— to describe systems for high- and low-g ground simulations, and our observations with animal cells under altered g conditions.

Here we discuss mainly experiments and technology dedicated to the study of the behavior in space of microorganisms and animal cells cultured in vitro. We will not discuss equipment for biomedical research, nor for growing plants in space. However, in Section 2 a we will give a survey of the major achievements of life science experiments in space.

Valuable results were obtained on automatic satellites. Unfortunately, complicated mechanisms are often subject to failures. Only the constant presence of man as an intelligent observer and operator can guarantee a critical evaluation of the experiment and a proper function of the equipment. Finally, experiments in space should be, whenever possible, preceded by ground simulation at high and low-g. We will describe a number of interesting results obtained by this approach.

2 Biology in Space

This section presents an outline of the topics belonging to the field of space biology and of their scientific background. The past and the future of space biology have been outlined in a brochure edited recently by Bjursted [1]. Although we will not discuss the technology of growing plants in space laboratories, it must be pointed out that plants are ideal objects for the study of gravitational effects since plant growth is regulated by well identified gravity receptor cells (statocytes) carrying gravity-sensitive organelles (statoliths).

2.1 Man in Space

The primary question at the beginning of the astronautic era was whether humans can survive at all for a prolonged period of time in space. This question was associated with the development of a life-supporting system (energy, atmosphere and waste disposal) capable of providing adequate living conditions to the crewmen. We can regard this phase as completed although technology will progress toward better and better solutions (see Sect. 9).

The physiological problems of man were identified already at the early times of astronautic and were sharply focused after the recent long-duration Soviet flights on the Salyut-6 station. A comprehensive discussion of adaptation of man in space is given in Ref. [2]. A brief summary is presented here.

a) The space motion sickness involves mainly the vestibular apparatus. Its occurrence is unpredictable and cannot be detected by ground simulations. It may be accompanied by serious symptoms of indisposition which can prevent astronauts from performing flight operations.

b) Cardiovascular changes: within few hours in weightlessness, approximately 2000 ml of body fluids (plasma and interstitial liquids) are shifted from the lower to the upper parts of the body. This effect does not cause serious inconveniences.

c) Degradation of bone material and muscle atrophy. These symptoms seem to be in part irreversible. Intense fitness training inflight does not prevent completely the diseases.

d) Hematological and immunological changes: A relevant loss of red blood cells, hemoglobin and plasma volume are observed after every mission. Lymphocyte reactivity is often reduced after spaceflight.

We will discuss here in more detail only those aspects of the adaptation of man in space which can be investigated by biotechnological applications such as *in vitro* simulations of cellular events.

Interesting examples are the immunological and hematological systems: Reduction of red blood cell mass (2—21%) and of hemoglobin mass (12—33%) is generally observed after the US and Soviet space mission. The changes are accompanied by a loss of plasma volume (4—16%) [3-9]. Erythrocyte and hemoglobin concentrations in the blood remain constant, suggesting that the changes are driven by a feed-back mechanism. Immunological changes consist mainly of reduced T-lymphocyte reactivity. The results of the 96-day and 140-day Salyut-6 missions suggest that the adaptation of the immune system to spaceflight occurs in two stages: The first takes place during the first 2—3 months in space, the second follows and consists of further weakening of the immune response [9 – 14].

It is important to point out that the RBC mass reduction and the depression of lymphocyte reactivity never harmed the health of the crews. The changes reflect physiological adaptation reactions rather than pathological conditions. However, with the advent of the Space Shuttle and of large space stations, the opportunity of flying in space will be offered to a broader community. The selection of crews will be less severe than in the past and therefore the hazard of anemic diseases and infections will be higher. The causes of the changes are not fully understood and further investigations are needed to explain these phenomena.

In addition to the medical examinations routinely performed, it will be useful to investigate systematically in vitro some of the important cellular processes which appear to be influenced by the space environment. This kind of experiments, when performed on space laboratories, will permit to discriminate between the effect of stress of spaceflight on the organism and the effect of $0 \times g$ on the biological system under investigation. This will deliver an important contribution to applied and basic research in space. As shown in Sect. 5.1.1 certain aspects of the immune system can be investigated in vitro.

2.2 Cell Biology

One of the most appealing features of experiments with living cells in weightlessness or at high-g is the transformation of gravity, a physical entity always constant in our ground laboratories, into a variable parameter like temperature or concentration. Consequently, living organisms which underwent evolution and development in a constant gravitational environment are suddenly confronted with a new situation. Therefore, the survival and proliferation of mammalian cells in altered gravitational fields is a challenging aspect of space biology.

The effect of varying the gravitational fields (mainly high-g) has been investigated on a variety of living organisms as long ago as 1806 (see Ref. [15] for a summary). The studies included plants, frog and sea urchin eggs, bacteria and amoeba, as well as complex organisms such as rat and man. Generally, the effects are more dramatic with increasing complexity of the investigated organism. A more detailed description of experiments with isolated cells in space is given in Sect. 3. Calculations made by Pollard [16] show that the distribution of cell organelles like mitochondria, nucleus, nucleolus and ribosomes may be influenced by gravity provided there is sufficient freedom of movement within the cell. This condition is satisfied in cells larger than 1 μm. Therefore, animal cells should be more subjected to gravitational effects than bacteria. In fact, when bacteria were grown in a centrifuge at $50.000 \times g$ [16] no effect was observed. However, the calculations of Pollard do not take into account cytoplasmic interactions like those involving cytoskeleton. In fact, intracellular movements are severely impaired by rather rigid structures. Gravity may interfere with cytoplasmic streaming as calculated by Kessler [17]. Folkman and Moscona [18] described the correlation between cell shape and growth: Cells of various lines in suspension are spherical whereas cells adhering to the walls of a culture flask are rather flat. It was found that when cells are converted from a flat shape into a spheroidal shape, cells incorporate less ^{3}H-thymidine into DNA. These findings provide further arguments in favor of experiments studying the effect of gravity on cell proliferation since one can expect that cell shape is influenced somehow by forces and tensions related to gravity.

However, Todd [19] detected no influence of gravity on the behaviour of the mitotic spindle in cultured human kidney cells. No effect of spaceflight was observed on human embryonic cells flown on Skylab 3 [15]. However, none of the experiments mentioned involved cell differentiation events.

Our findings with lymphocytes cultured at different g-levels are described in Sect. 8. Briefly, high gravity has a stimulatory, simulated low gravity an inhibitory effect on lymphocyte activation by mitogens.

2.3 Radiobiology

Several kinds of radiations were detected in space ranging from UV and X-rays to high-energy particles. Among the high-energy particles (electrons, protons and heavy ions) the most important for radiobiologists are the HZE (high-charge and high-energy) particles, mainly iron nuclei. Although astronauts were hit by HZE particles, registered as flashes of light, during the past missions, no consequences for their health has been reported. Calculations for a standard Space Shuttle/Spacelab orbit show that flux of Fe nuclei is approx. 2.5 Fe cm^{-2} sr per day at solar maximum activity and 0.8 Fe cm^{-2} sr per day at solar minimum activity. About 10—20% of Fe ions have energies less than 500 MeV/nucleon, and are highly ionizing. About 25% of Fe nuclei will interact with an aluminium wall of 5 g cm^{-2} thickness. This is the minimum wall thickness encountered by a particle penetrating into the Spacelab.

Particular caution should be taken in case of solar flares: the radiation consists of 95% protons. Surface doses may be higher than 1000 rads; however, the penetration of the radiation is quite low. The effect of cosmic radiation on biological objects was studied on Biosatellite II, on Apollo 16/17 and on the ASTP missions (see Sect. 3.1). Cell damage like cell death, tumor induction and genetic mutations may occur at different levels, depending on the organelles hit.

Within certain limits, the study of cosmic radiation effects can also be performed on the ground with the accelerators now available. However, a combined effect of radiation and microgravity can be achieved only in space laboratories. Finally, every biological experiment in space should take into account the effect of radiation. Dosimeters will record radiation in different locations on Spacelab.

2.4 Exobiology

The objectives of this discipline are to study origin of life in the universe and the detection of extraterrestrial life. Simulation experiments with primordial elements like hydrogen, oxygen and nitrogen showed that simple molecules (methane, water, ammonia) can be formed in a primitive atmosphere. These molecules can react with one another and produce amino acids, purines, pyrimidines and carbohydrates, which are the essential constituents of the molecules of life. The first and only attempt to search for extraterrestrial life through a biological approach was performed on the Viking missions in March 1975 (see Sect. 3.4). It may be of interest to the reader to know that the total number of technological civilisations which have appeared over the entire history of our galaxy has been estimated to be around a billion.

Table 1. US Missions carrying biological payloads

Mission	Date	Duration	Biological specimens	Hardware	Comments
Discoverer XVII	1960	3 d	Human γ-globulins, rabbit antiserum	Millipore filters, nuclear emulsion, tracking paper	Greater Ag/Ab reactivity
			Human conjuctival and synovial cells	Culture chambers, refrigeration units, dosimeter	Failed: Medium exhausted
			Chlostridium sporogenes, *Chlorella ellipsoidea*	Glass ampoules	No effect of spaceflight
Discoverer XVIII	1960	3 d	Human amnion, conjunctival, sternal marrow, synovial monocytic, leukemia, HeLa, embryonic lung, and chicken embryonic cells	Glass ampoules, refrigeration unit, filmpack	Failed due to unfavorable experimental conditions
			Clostridium sporogenes	Glass ampoules	No effect of spaceflight
			Neurospora crassa	Millipore filters, photoemulsion	No effect of spaceflight
Discoverer XXXII	1960	3 d	Corn seeds (*Zea mays*)	Container	No effect of spaceflight
Mercury 5	1961	34 h	Chimpanzee	Life supporting system, sensors, photographic equipment	No effect of spaceflight, animal survived without damage
Gemini 3	1965	5 h	Whole human blood	Aluminium holder, ^{32}P, dosimeter	No effect of spaceflight
			Sea urchin eggs	Container with fixative operated through-handle	Failed for mechanical reasons
Gemini 8	1966	11 h	Frog eggs	Chamber with holder for chemical fixation	No effect of spaceflight
Gemini 11	1966	3 d	*Neurospora crassa*, whole human blood	Aluminium holder, ^{32}P, dosimeter	No effect of spaceflight
Gemini 12	1966	4 d	Frog eggs		
Biosatellite I	1966	—	Several biological experiments	Same experiment as on Gemini 8	Spacecraft did not react on command and burned on reentry after 60 d on orbit
Biosatellite II	1967	2 d			Part of the scientific yield reduced by the early reentry of the spacecraft

Mission	Duration	Specimens	Equipment	Observations
		Amoeba, *Artemia salina*, wheat seedlings, *Saccharomyces cerevisiae*, *Neurospora crassa*, *E. coli*	Package, thermistors, dosimeters, cytospectrophotometer, photocameras	No effect of spaceflight
		Wasps		Change in behavior Greater amino acid changes, response to $0 \times g$ similar to that in clinostat
		Pepper plant		Disturbed spindle function of root tips Faster growth
		Tradescantia	Acrylic modules, thermistors	No effect of spaceflight
		Salmonella typhimurium Frog eggs *Drosophila*	Package, ^{85}Sr	Inconclusive results due to contamination of capsule with chemicals. $0 \times g$ and radiation cause premature aging and chromosome damage
		Tribolium confusum	Housing compartment, ^{85}Sr, dosimeter	Abnormalities due either to $0 \times g$ and radiation or to temperature drop during flight operations
Biosatellite III 1969	8.5 d	Pig-tailed monkey	Life-supporting system, several sensors	Early call down due to deterioration of physiological conditions. Death 8 h after recovery due to dehydration and electrolytic imbalance
OFO-1(A) 1970	6 d	Bull frog	Animal immersed in water, sensors, life-supporting system	Only partial adaptation of the behavior to $0 \times g$
Apollo 14 1971	9 d	DNA, hemoglobin, dyes	Zonal electrophoresis equipment	Feasibility of electrophoresis at $0 \times g$ demonstrated by separation of dyes
Apollo 16 1972	11 d	*Drosophila* *Artemia salina* and grasshoppers eggs, *Tribolium confusum*, *Arabidopsis thaliana* seeds, bean embryos, *Bacillus subtilis*, protozoan cyst	Package, ^{85}Sr Biostack I	No effect of spaceflight Discussed in detail

Table 1. (continued)

Mission	Date	Duration	Biological Specimens	Hardware	Comments
			Nematode larvae, *Bacillus thuringiensis*, *Bacillus subtilis* (4 strains), *E. coli* T-7 phage, *Chaetomium globosum*, *Trychophyton terrestre*, *Rhodotorula rubra*, *Saccharomyces cerevisiae*	Microbial ecology evaluation device with 798 cuvettes, dosimeter, thermometer	No effect of spaceflight
Apollo 17	1972	12 d	Biostack II: identical to biostack I Pocket mice	Cannister, life-supporting system, dosimeters implanted in brain	Four mice of five returned alive, 80 HZE events recorded, no significant lesion found
Skylab 2	1973	28 d	*Bacillus subtilis*, *Bacillus mycoides*, *E. coli*	Petri dishes, incubator, photocamera	Faster growth, fewer but larger colonies, morphological changes, more sensitive to antibiotics
Skylab 3	1973	59 d	Antigens/antibodies	Immunodiffusion plates, thermos	No effect on Ag/Ab reaction
			Human embryonic lung cells	Incubator	Described in detail, no effect of spaceflight
			Elodea (water weed)	Vials, microscope, picture camera	Failed, plants died
			Rice seedlings	Package, camera	Irregular growth
			Cross spiders	Cage, photoequipment	Spiders use gravity-sensitive organ for web formation
			Killifish	Polyethylene bags, photoequipment	$0 \times g$ acts as vestibular stimulus: swimming anomaly
Skylab 4	1973/74	84 d	*Elodea, Bacillus subtilis, Bacillus mycoides, E. coli*	Petri dishes, incubator, photocamera	Elodea died, the other microorganisms showed the same changes as on Skylab 2
ASTP (Apollo-Soyuz Test Project)	1975	9 d	Human, horse, rabbit RBC, human lymphocytes and kidney cells, rat bone marrow, spleen and lymphnode cells	Zone and free-flow electrophoresis, isotachophoresis equipment	Separation and concentration of cells achieved, discussed in detail
			Streptomyces levoris	Petri dishes, incubator, photocamera	Decreased growth rate periodicity, no changes in biorythm

		Artemia salina, Tribolium confusum, *Caraustus morosus,* tobacco seeds, *Arabidopsis thaliana, Zea mays,* *Bacillus subtilis*	Biostack III	Discussed in detail
		Killifish embryos	Polyethylene bags, photoequipment	No effect on embryonic development
Viking	1976	Mars soil	Fully automated equipment for search of primitive life on Mars	Discussed in detail

3 Biological Payloads on US Missions

In this section experiments performed aboard manned and unmanned US space stations are reviewed. As pointed out before, very little research has been done aboard manned satellites. A number of interesting experiments were carried out on automated satellites, mainly on the US Biosatellite II. However, some projects failed for mechanical reasons.

We discuss here in detail those projects which are most relevant to space biotechnology. A concise summary of all biological experiments carried out on US missions has been recently published by NASA [20].

A consistent program of basic and applied research in life science will start when the Spacelab becomes operational. There, facilities like incubators, centrifuges, microscopes, freezers and refrigerators will offer to the user acceptable working conditions (see Sect. 5). In addition, a selected group of scientists is presently trained by NASA as mission and payload specialists for biological experiments. However, we must always keep in mind that safety requirements and constraints like weight, energy and crew time will always limit the scientific goals of the investigations.

Two aspects are relevant: a) The instrumentation carried by the space vessel, and b) the effects of the space environment on growth and behaviour of the organism studied. Table 1 compiles the biological experiments carried on US missions so far [20]. Biomedical investigations on humans or experiments with animals or plants are not discussed. The instrumentation employed was, except for few cases, very simple and mainly consisted of passive containers (petri dishes or sealed ampoules) in which the microorganisms were kept in a nutrient medium. The temperature control was rather primitive and mainly consisted of thermos containers. Man power invested in biological experiments was very limited or nil. The duration of the missions was in most instances too short for an extensive study of the effect of spaceflight on growth and survival of microorganisms. The interest of the investigators was directed toward the detection of radiation damages rather than toward the effect of weightlessness per se. Two of the most important missions, namely Biosatellite I and II either failed or landed too early, upsetting the schedule of most experiments. Therefore, it would be premature to draw conclusions on the adaptation of living systems in space from the studies which have hitherto been performed. We should use the experiences gained for a better choice of the system to investigate and for improving the reliability of the equipment. More biochemical parameters should be analyzed in the future. DNA, RNA and protein biosynthesis are useful indicators of intracellular changes still unexplored in weightlessness. Table 1 shows that very little has been done in this direction particulary with animal cells. In addition, differentiation processes have never been studied in vitro. Lymphocytes and hemopoietic cells are good test objects for this purpose (Sect. 5.1.1). Four relevant projects performed on US missions are discussed here in more detail.

3.1 Biostack Experiments

Biostack was one of the first sophisticated experimental devices carried out in space. It consisted of a passive container which did not require power or crew interface. Biostacks I, II and III were flown on the Apollo 16, 17 and Apollo-Soyuz Test Project

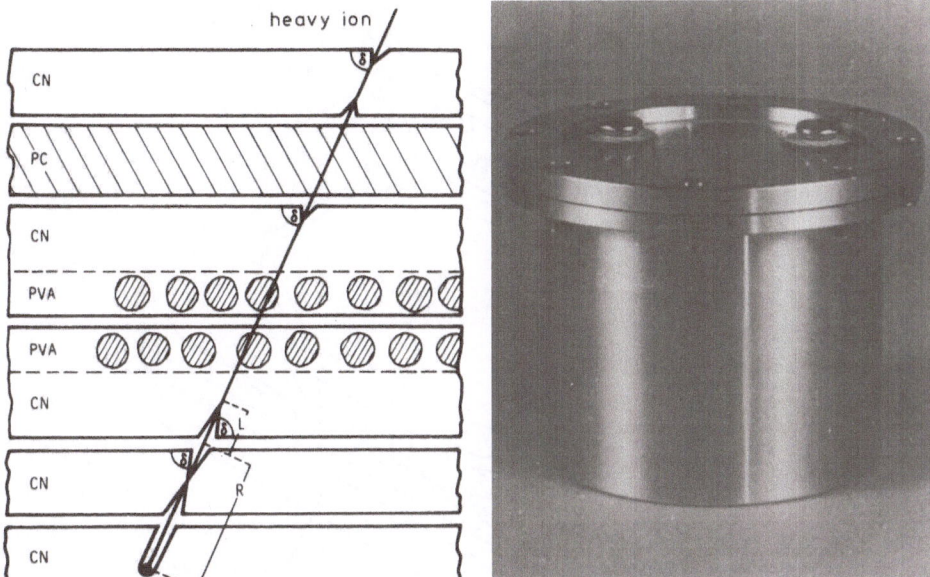

Fig. 1. Schematic representation of the biostack conception. Left: Monolayers of biological objects are sandwiched alternately with nuclear track detector sheets. Right: Biostack flight container. CN: cellulose nitrate, PC: polycarbonate, PVA: poly-(vinyl alcohol) (courtesy of H. Bücker, DFVLR, Frankfurt)

(ASTP) missions, respectively. The biostack experiments were designed and evaluated by a group of 30—40 investigators in Europe and USA [20]. The objective of these experiments was to study the effect of cosmic radiation on a variety of biological objects (Table 1): Microorganisms, eggs and plants seeds were embedded in poly(vinyl alcohols) between photographic emulsion layers (Fig. 1). This concept first proposed by Eugster [21] and developed by Bücker [22] allows the identification of the object hit by the radiation and to track the path of the penetrating particle through the biological object.

Eggs and plant seeds were grown after landing in the ground laboratories. Several effects of radiation were detected. Most interesting are *Artemia salina* eggs developed to individuums with abnormalities in the extremities, in the abdomen and in the thorax. *Zea mays* seeds produced plants with leaves showing large unusual yellow strips.

Biostack is a typical example of a simple, passive equipment without energy or crew operation requirements with a high scientific yield. However, the equipment is suitable only for investigations of the effect of radiation whereas gravitational effects cannot be detected with this approach.

3.2 Tissue Culture Incubator on Skylab

The effect of spaceflight on human embryonic lung cells WI-38 was studied during the 56-day Skylab 3 mission by Montgomery et al. [15,23]. The experiment was

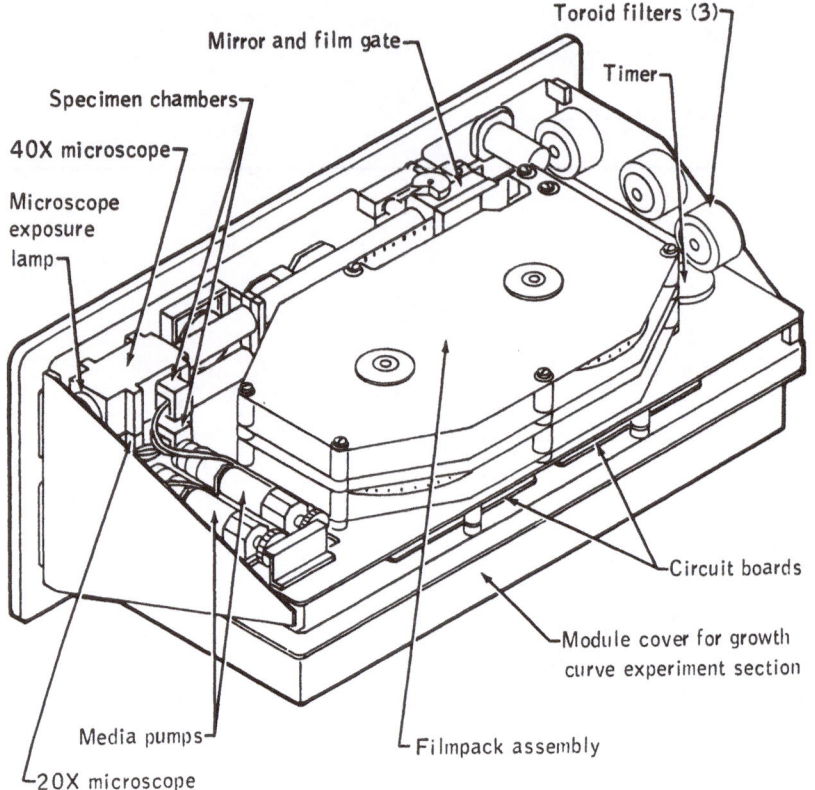

Toroid filters (3)

Mirror and film gate

Timer

Specimen chambers

40X microscope

Microscope
exposure
lamp

Circuit boards

Module cover for growth
curve experiment section

Media pumps

Filmpack assembly

20X microscope

Fig. 2. Internal set-up of Woodlawn Wanderer 9 incubator (courtesy of R. G. Thirolf)

performed with one of the most sophisticated equipment ever used in space biology[24]. It consisted of a fully automated tissue culture package called Woodlawn Wanderer 9, designed to achieve four major objectives:

— To keep cells alive in culture by supplying fresh medium at 36 °C for several days;
— to record phase contrasted pictures with a time-lapse camera for 28 days;
— to fix several specimens of cells at given times;
— to recover living cells after flight for further culturing in the ground laboratory.

The instrument weighed 10 kg, measured $40 \times 19 \times 17$ cm and was powered by 28 DC with an average consumption of 16 W at 10 °C room temperature. The package was separated into two main compartments: A camera-microscope section and a redundantly sealed growth-curve experiment section (Fig. 2).

The camera-microscope section consisted of two independent 20 and 40-power *camera microscopes*. The pictures were recorded on a 16 mm microfilm. The cells were grown on glass in perfusion chambers filled with cultures containing 7000 cells/ml. After the cells became attached to the lower glass disk, the chamber was fixed in the microscope and focused. Each chamber had a volume of 105 µl and was connected to an automatic perfusion system adding fresh medium every 12 hours.

The growth-curve experiment section was contained in a module easily removable for biological servicing and consisted of two separated identical and independent

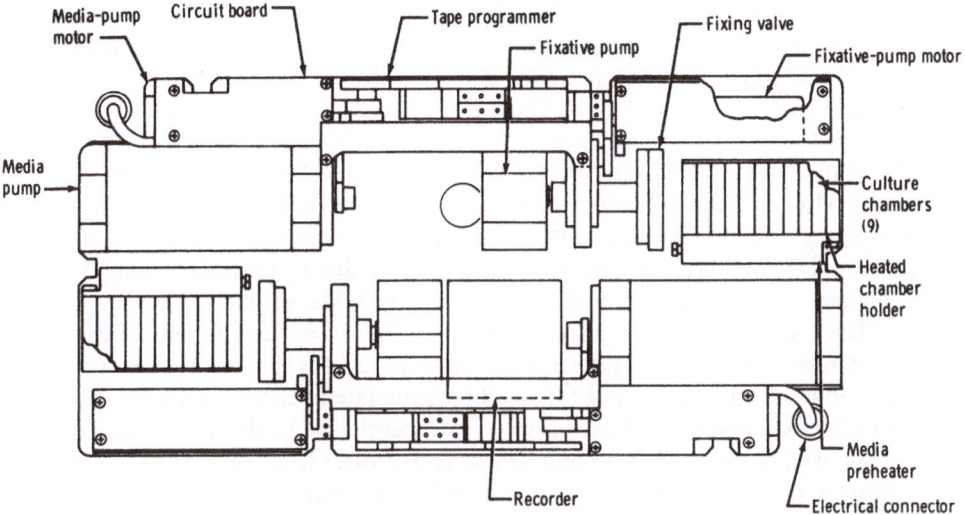

Fig. 3. Growth-curve experiment section of Woodlawn Wanderer incubator (courtesy of R. G. Thirolf)

units (Fig. 3). Each unit was composed of nine cell culture chambers. Here too cells were fed automatically. At pre-programmed times, a fixative (5% glutaraldehyde in Earle's balanced salt solution) was added. On mission day 12, four of the nine chambers were not fixed but maintained at room temperature for the rest of the mission. The cells were recovered alive and further cultured in the ground laboratory. An extensive analysis of the collected data showed that the space environment did not produce detectable effects on WI-38 embryonic cells. The technology of this experiment is the result of a remarkable effort performed by scientists and NASA engineers. However, the funds for this sophisticated equipment were granted at a time of great euphoria of spaceflight. Nowadays, it is unlikely that any national or iternational agency will commit itself to financing an expensive equipment for only one experiment. In addition, WI-38 cells were attached to a substrate; therefore, possible effects of $0 \times g$ on a dynamic cell-cell interaction as well as specific differentiation processes which are not taking place in WI-38 cells could not be investigated. Finally, cells anchored to a substrate forcefully assume a flat configuration regardless of the g-environment. As pointed out above (Sect. 2.2), a correlation has been found between growth and cell shape. Cell shape itself can be influenced by the g-environment. In conclusion, there are reasons indicating that cells growing in suspension and the possibility to induce them to differentiate are better objects for investigations in space.

3.3 Separation of Cells by Electrophoresis

Bioprocessing of living cells is conditioned by the availability of homogeneous cell populations. The separation of specialized animal cells is one of the challenging problems of biotechnology. Since the electric charge localized on the membrane is a characteristic property of each cell, it is, in theory, possible to purify cell populations by means of electrophoretic techniques. However, cells in suspension rapidly sediment to the bottom of the culture chamber and, on the other hand, Joule effects generate turbulences in the separation chambers by thermal convection. The disadvantages of thermal convection and sedimentation are ruled out in a weightless environment. Therefore, preliminary attempts to achieve separation of biomaterials by electrophoresis at $0 \times g$ were performed in the Apollo 14 and 16 missions [25].

On Apollo 14, electrophoretic separation was successfully achieved for the first time in space with a mixture of DNA, hemoglobin and dyes. In spite of a number of problems encountered during operations, mainly due to bacterial contamination and too long storage, separation of dyes was achieved. The second attempt on Apollo 16 was less successful since the electrophoretic column leaked. The sample consisted of polystyrene latex beads of different sizes as a model for living cells. The scientific and technological importance of this method stimulated further development of the electrophoresis equipment: Two projects, one developed in the USA, the other in Germany, were flown on the Apollo Soyuz Test Project (ASTP) in 1975. In the two following sections, the experiments are described in detail.

3.3.1 Static Discontinuous Electrophoresis

This experiment was developed in the USA and carried the number MA-011. R. E. Allen was the principal investigator [26, 27]. It was decided to reuse those parts of the Apollo 14 and 16 experiments which worked correctly and to change the design of those parts which failed. Bacterial degradation was avoided by using sterile procedures. Two techniques were tested: zone and isotachophoresis. The biological samples were selected according to the following criteria:

a) Glutaraldehyde-fixed rabbit, human and horse red blood cells (RBC) provided a material resistant to mechanical stress and to hemolytic agents. The bands of separated RBC are easily distinguished on photographs.

b) Human lymphocytes offer an ideal system consisting of several subpopulations. The separation of immunoglobulin-producing B-lymphocytes from lymphokines-secreting T-lymphocytes is a very important task for immunologists and cell biologists. In addition, a variety of T-lymphocyte subpopulations has been recently identified. The electrophoretic separation of lymphocytes, together with immunological techniques are the most promising approaches since other separation procedures such as sedimentation velocity are not applicable due to the almost identical morphology, density and size of lymphocyte subpopulations.

c) A third sample contained human kidney cells. The separation of urokinase(UK)-producing cells would be a very important biotechnological achievement. Urokinase is an enzyme converting plasminogen to plasmin, an obligatory step in the lysis of blood clots. Therefore, UK is an important thrombin-preventing agent. The demand for UK exceeds production capacity. UK is produced in vitro by human fetal kidney cells. However, only 5% of the cells isolated from the cortex of the kidney by trypsinization produce UK. Consequently, the production costs of UK can be reduced by isolating and culturing UK-producing cell subpopulations. As explained above, electrophoresis could provide optimal conditions for a successful separation.

The electrophoretic unit flown on the ASTP mission weighed 13.61 kg, measured $30.3 \times 20.3 \times 40.6$ cm. By means of a self-contained power supply, requiring an external

208-A AC, three phases and 400 Hz connection, the voltage could be varied between 0 and 99 V. Two current levels could be selected at 4.0 and 1.31 mA. A thermoelectric unit controlled the temperature between −40 and +5 °C with the purpose of cooling or freezing the electrophoretic columns. The columns were illuminated by fluorescent light during the experiment and the evolved gas was removed by a phase separator. The separation of RBC bands was recorded by a 70 mm electric camera. The eight electrophoretic columns consisted of pyrex glass tubes of 0.635 cm inside diameter and 15.25 cm length. They were split lengthwise and rejoined by a silicone seam to compensate the expansion of the buffer solution during freezing after electrophoresis. The biological samples were contained as frozen slides in 0.13 mm thick tetrafluoro-ethylene cover. The slides were stowed before launch in a cryogenic canister and cooled with liquid nitrogen. They were removed immediately before the experiment was started and inserted in the separation column.

The phosphate buffer (pH 7.3) was supplied by a pump. After the separation was completed the columns were frozen by operating the thermal control unit and subsequently stowed in the freezer until recovery. The inflight operations required minimum crew operations.

In the ground laboratory, the frozen buffer containing the processed samples was separated from the column and sliced in 0.5—5 mm long sections.

The lay-out of the samples was the following: Columns 1, 2, 3, 5, 6, 7 were used for zone electrophoresis, 1 and 5 contained fixed rabbit, human and horse RBC, 2 and 6 human lympocytes, and 3 and 7 human kidney cells. Columns 4 and 8 were used for isotachophoresis and contained fresh rabbit and human RBC, fixed rabbit and human RBC, respectively.

The equipment met with expectations since electrophoresis in a space environment has been demonstrated to be feasible. However, the fluid line of columns 2, 5, 6 were clogged during pre-flight assembling causing accumulation of chemical and gaseous products in the electrode region; this resulting in voltage and pH fluctuations. In addition, column 1 was damaged during the slicing procedure. The photographs taken in the flight during the experiments showed that separation of RBC was achieved satisfactorily in column 1. Unfortunately, both columns with the lymphocyte probes failed. However, subsequent analyses revealed that the experimental conditions did not affect their viability, morphology and ability to secrete products. This is a very encouraging observation in view of future applications.

The separation of kidney cells in columns 3 and 7 gave the most satisfactory results. Thus, UK-producing cells accumulated in one band and were recovered viable. This success will undoubtedly stimulate further technological developments in space laboratories.

For various reasons, the isotachophoretic experiments with columns 4 and 8 did not run for a sufficient time to permit visualization of intercompartmental boundaries. However, it could be shown that this approach offers the advantages of sharp boundaries and of enrichment of the migrating zones. Proper spacers have to be developed for the fractionation of living cells. In conclusion, the MA-011 experiments demonstrated the applicability and the great advantage of electrophoretic techniques in space for basic and applied research. An improved equipment to be flown aboard the Space Shuttle is presently under development in the biotechnological units of NASA at Johnson Space Center in Houston (see Sect. 6).

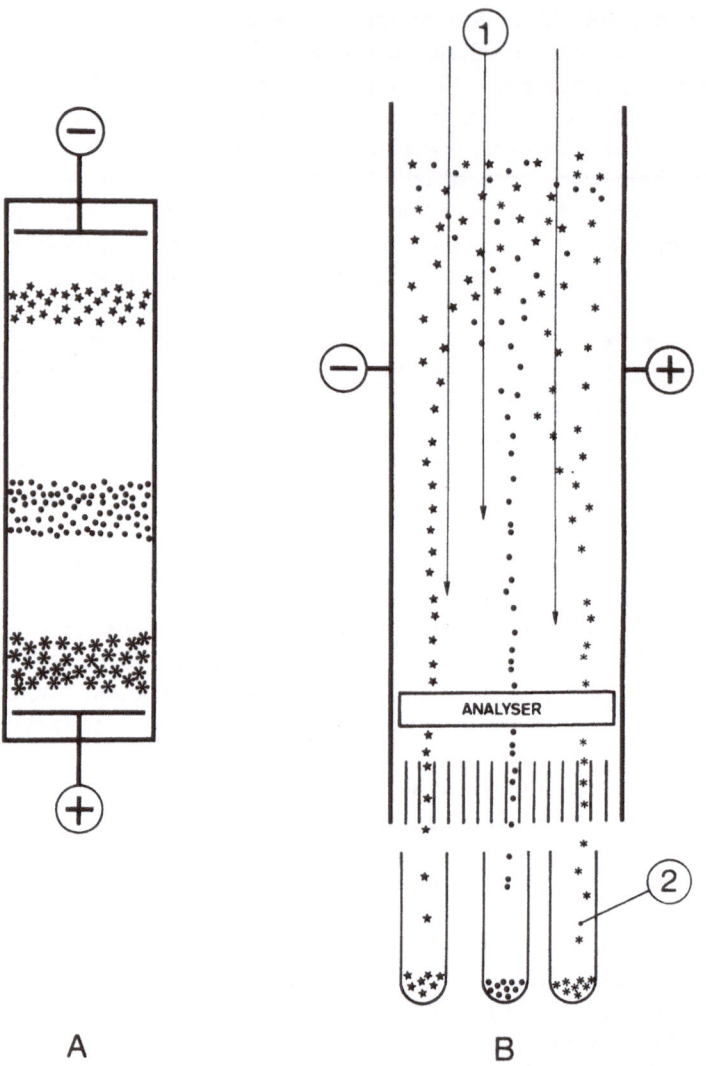

Fig. 4. Principle of static, discontinuous (A) and of free-flow continuous electrophoresis (B), 1: Flow direction of buffer with samples, 2: Collection of fractions

3.3.2 Free-flow Continuous Electrophoresis

The principle of the technique is shown in Fig. 4. The great advantage of free-flow electrophoresis is that theoretically an unlimited amount of material can be continuously processed. In addition, several samples can be easily inserted in sequence for separation. The disadvantage is an inevitable band broadening at $1 \times g$ and at $0 \times g$ as well.

The heart of the apparatus developed by Hannig [28, 29] is the separation chamber consisting of two parallel plates isolated by a gap of 4 mm (Fig. 5). The electric field

Fig. 5. Free-flow continuous electrophoresis equipment flown on the Apollo-Soyuz Test Project (from Ref. [49])

was generated by 180 mm long electrodes on the sides. The electrodes were shielded from the buffer in the chamber by ion exchange membranes. The electrode compartments were purged from gases by streaming buffer driven by a peristaltic pump. The gases generated in the buffer solution were adsorbed onto palladium. During operation, the plates were cooled at 5 °C. The buffer was driven by another peristaltic pump from the container through the separation chamber upstream the sample inlet. In this way, a constant laminar flow was produced. The buffer and the sample entered a waste container after passing the separation chamber. The samples were stored in a refrigerator at 0—5 °C. The separation was evaluated by an opto-electronic device. Totally, 4 samples were run: a) Rat bone-marrow cells, b) human and rabbit RBC, c) rat spleen cells, d) rat lymph-node cells supplemented by human RBC as a marker.

A satisfactory evaluation of the separation was hindered by a malfunction of the light for optical detection. Only the zones with maximum absorption could be detected by the optoelectronic device. Nevertheless, the difficulties could be circumvented and satisfactory separation curves could be obtained.

The results show that with the bone marrow cells an excellent separation was achieved. The resolution was much sharper than the corresponding separation realized in a control experiment on the ground. Conversely, insufficient information was available from the second sample, a mixture of human and rabbit RBC. Spleen cells gave the best result; several cell populations were separated and concentrated in sharp bands. In the case of sample number 4, a good resolution of lymph-mode cells was achieved at $0 \times g$ as well as in the $1 \times g$ ground control.

Again, experiment MA-014 demonstrated the advantages of separating animal cells by electrophoretic techniques at $0 \times g$. A further development of the apparatus described here is being prepared by Hannig to be used aboard the Spacelab (see Sect. 5.2.3).

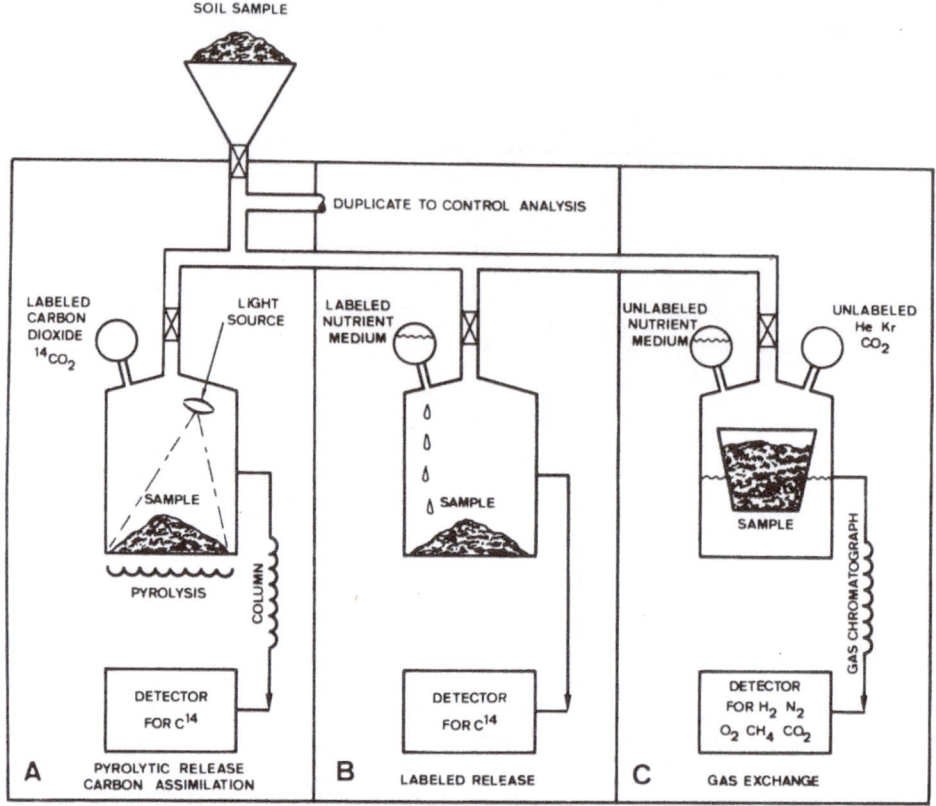

Fig. 6. Diagram of the Viking experiment searching for primitive life forms on Mars. Samples of Martian soil were tested for (from Ref. [30]):
A: Synthesis of organic matter from labeled CO_2,
B: release of labeled gases from labeled nutrients,
C: change of the gas composition by metabolic action.

3.4 The Viking Mission to Mars

After travelling nearly one year through the solar system, Viking 1 set down on Mars surface on July 20, 1976, followed by Viking 2 on September 3, 1976.

In addition to instruments for soil and atmosphere analysis, Viking carried a sophisticated automatic miniaturized laboratory for search of primitive life forms on the Martian soil [30-33]. Three independent biological experiments were performed on soil samples:

a) Pyrolytic release experiment (Fig. 6A): The sample was incubated for five days in a chamber under martian conditions in the presence of ^{14}C-labeled carbon dioxide and carbon monoxide. Artificial sunlight for photosynthesis was provided by a lamp. After incubation, all organic materials were pyrolyzed in two steps at 625 and 700 °C. Radioactivity was measured by a detector. In a control experiment martian soil was first sterilized at 170 °C prior to incubation and pyrolysis.

b) Label-release experiment (Fig. 6B). Soil samples were incubated in the presence of nutrient medium containing ^{14}C-labeled formate, lactate, glycine, alanine, and glycolic acids. Living microorganisms, when present, would release labeled carbon monoxide and carbon dioxide or methane. Again, a control sample was sterilized prior to incubation.

c) Gas-exchange experiment (Fig. 6C). Martian soil was incubated with unlabeled medium in a controlled atmosphere of He, Kr and CO_2 for 7 days. Gases like hydrogen, nitrogen, oxygen, methane, and carbon dioxide eventually released by microorganisms were measured with a conductivity detector.

Initially, the experiments delivered results which were consistent with the presence of life forms on the martian surface. However, the chemical analysis of soil samples revealed no organic compounds larger than methane and propane [1]. A possible explanation of the puzzling results is that the reactions observed may be catalyzed by peroxides (γ-Fe_2O_3) present in the surface of the planet.

In conclusion, this exciting project did not give a definite answer to the question whether there is life on Mars or not. Further experiments which must take into account the peculiar composition of the Martian soil are required to solve the problems. However, simulation experiments conducted on earth make a biological interpretation of the Viking results most unlikely [34, 35].

4 Biological Payloads on Soviet Missions

The material presented in this section has been selected from: a) translations into English from Russian of original articles, memoranda and bulletins, mainly from the Journal of Space Biology and Aerospace Medicine which is a translation (edited by NASA) of Kosmicheskaya Biologiya I Aviakosmicheskaya Medsina, Moscow; b) original reports published by the Soviets mainly in the COSPAR (Committee on Space Research) series Life Sciences and Space Research. A comprehensive collection of literature references concerning soviet experiments appeared recently [36]. A great number of biological experiments was performed on manned and unmanned Soviet satellites. Unfortunately, very few technical details of the hardware have been published. The Soviet space program in life seciences was concerned with the study of adaptation of man in space, basic research of the effects of $0 \times g$ and cosmic radiation on microorganisms, plants, seeds and seedlings, embryos, mammals (mainly rats) and, interestingly, with the development of autonomous greenhouses in space. We can divide the activities of the Soviets in three phases:

a) The pioneering phase of manned spaceflight, characterized by the missions of the spaceships Vostock I—VI and Voskhod I and II (1961—65). Already cosmonaut Gagarin carried on Vostock I containers with E. coli and other microorganisms [37]

b) The second phase is marked by the launch of automated biosatellites of the series Cosmos (Table 2) containing highly sophisticated equipment like life-supporting systems for rats and centrifuges generating $1 \times g$ conditions.

c) The era of a systematic exploitation of space resources began for the Soviets in 1971 with the launch of orbital stations periodically visited by crews travelling on Soyuz vessels and refurbished by unmanned space freighters of the type Progress. The launch of the Salyut-1 space station was followed by Salyut-3 (1974), Salyut-4

Table 2. Payloads on cosmos biosatellites[a]

Flight	Date	Duration	Biological specimens	Hardware	Comments
Cosmos 110	1966	22 d	Two dogs	Life-supporting system	Extensive changes of various physiological systems [38, 39]
			Cabbage and lettuce seeds, onion sets	Containers	Plants were grown after recovery: Less dry substances, higher ascorbic acid content and sugar biosynthesis in flight plants [40]
			Chlorella ellipsoidea, E. coli		No effect of spaceflight [37]
Cosmos 605	1973	22 d	Rats	Life-supporting system, γ-radiation source	Enhancement of radiobiological effects was observed in rats exposed simultaneously to γ-rays and to $0 \times g$ [41, 42]
Cosmos 613	1973	61 d	Crepis capillaris seeds	γ-Radiation source	Higher radiation damage at $0 \times g$ than in control tests [43]
Cosmos 690	1974	22 d	Rats	Life-supporting system, ^{137}Cs radiation source	Enhancement of radiation damage at $0 \times g$ [41, 42, 44]
			Polyporus brumalis		Abnormal growth [45]
Cosmos 782	1975	19.5 d	Rats	Life-supporting system, $1 \times g$ centrifuge	Effects of $0 \times g$ attenuated by artificial $1 \times g$ [42, 46]
			Drosophila		No effect of spaceflight [11, 47]
			Lettuce and tobacco seeds	Biobloc equipment	Chromosomal aberrations, decreased germination rate and abnormalities, changes also in objects not hit by HZE particles [48]
			Artemia salina eggs	Biobloc	Late developmental inhibition [49]
			Carrot tumor tissue		No effect of $0 \times g$, larger tumors in samples grown in artificial $1 \times g$ [47]
			Carrot embryo cells		No effect of $0 \times g$ [47]
			Guppies, killifish eggs		No effect on development [47]
			Bacteria, fungi, animal cells, tortoises		Results not yet available

Cosmos 936	1977	18.5 d	Rats	Life-supporting system, two 1 × g centrifuges	Artificial gravity attenuates physiological effects of 0 × g on bones, blood, organs, vestibular system etc. [42,46,50]
			Drosophila: Larval cultures and mature flies; Crepis capillaris, Zea mays	New transportable field laboratory at recovery site	No effect on development, reduced vitality [50-52]; Grown inflight: Cell ultrastructual changes, e.g. swollen mitochondria [51,52]
			Phycomyces blackesleamus		No effect of spaceflight [51,52]
Cosmos 1129	1979	19 d	Rats	Two 1 × g centrifuges, life-supporting system, field laboratory at recovery site	Results analogous to those of Cosmos 936 [42,53]
			Drosophila; Japanese quail eggs; Lettuce seeds		Grown between 0 and 1 × g [11,53]; Normal embryogenesis [53]; Chromosomal changes [53]

A further Cosmos Biosatellite carrying a primate is planned for 1982 [11,53].

ᵃ Biological samples were carried also on the Cosmos satellites 212 (1968) and 368 (1970) [36].

Table 3. Biological payloads on Soyuz/Salyut space stations[a]

Station	Launch-date	Biological specimens	Hardware	Comments
Salyut-4	1974	Drosophila	Biotherm incubator	Mutations observed in progenies [55,56]
		Dwarf peas	Growth on ion-exchange resin	After initial growth the plants died [55]
Soyuz 19	1975	Chlamidomonas reinhardi, Arabidopsis taliana	Biocat thermostat	Decreased survival [57]
		Crepis capillaris	Growth chamber	Chromosomal aberrations [57]
		Proteus vulgaris		Several biochemical changes, morphological changes at cellular level [57]
		Fish: Brachydanio rerio		No effect of spaceflight [57]
Soyuz 20	1975	Gladiolus muscari racemosus		Recovered after 92 d in space, blossomed earlier than in control tests and gave blossoms of a different color maintained through generations [11]
		Anethum graveolens seeds		Inhibition of germination
Salyut-5	1976	Roe corn of fish Brachydanio rerio	Biofixator at 23 °C	Roe corn is cultivated and subsequently the larvae are fixed after 8–10 d at $0 \times g$. Electron microscopic analysis of the ultrastructure of the vestibular apparatus shows no effect of $0 \times g$ [58]
		Lebistes reticulatus: Adult male and female fish	Aquarium with oxigenated water	The air formed gas bubbles in the water. After some days at $0 \times g$ the fishes swam with their back fin oriented toward the bubble [58]
		Polyporus brumalis	Geotrop	Normal fungal development [58]
		Crepis capillaris seeds and seedlings, Arabidopsis thaliana seeds	Germination in Biotherm incubator [56] at 24 °C	Samples for 18 to 249 d in space: Increase in chromosomal aberrations. $0 \times g$, dynamic factors and cosmic radiation can modify the radiation effect of additional γ-irradiation [58]
Salyut-6	1977	Chinese hamster cells	Cytos incubator	Data on morphology, growth and functional activity not yet available [11]
		Quail eggs		Differences on embryonic development [11]
		Paramecium aurelia		Faster growth in space [59,60]
		Seeds of Arabidopsis, onions, radishes, cucumbers, lettuce, peas, seedlings of borage, fennel, parsley, dill, garlic, foliage, tulip bulbs	Greenhouse and Gravitat providing artificial gravity	Plants are incapable of performing a full growth cycle at $0 \times g$. Under artificial gravity conditions the development is normal and photosynthesis is not disturbed [61,62]

Interferon	Ampoules	Packed in liophilized, gel and liquid form [63, 64]
		Effect of radiation studied [63, 64]
Drosophilla, Chlorella		
Amphibians	Emkon-T aquarium	Frog eggs were fertilized before launch [64]

a This table records only a selection of biological experiments. A comprehensive collection of literature references has been recently published by NASA [36]. Soyuz 19 docked with an American Apollo spacecraft during the ASTP mission. Soyuz 20 docked with the Salyut-4 station. The Salyut space laboratories were repeatedly visited by crews travelling on Soyuz spaceships.

(1975), Salyut-5 (1976), and Salyut-6 (1977). Salyut-6 is still operational in 1981 and hosted two spectacular missions lasting 6 months in 1979 and 1980.

In the decade 1961—70 [37], a number of biological payloads were carried aboard manned and unmanned satellites like Satellite-ships 2, 4, 5, Vostock 1—6, Voskhod 1, 2 and Zond 5—8. The payloads included *E. coli* B and K-12(λ)$^+$, *Aerobacter aerogenes* 1321, *Staphylococcus aureus* 0—15, *Clostridium butiricum*, yeasts, *Chlorella*, Tobacco mosaic and Flu viruses. Generally speaking, no effect of spaceflight was observed on growth, development, cell and nuclear division, and on mutagenesis. The materials were carried in simple containers, and no particular experiment or manipulation were performed during flight. A small, but statistically significant phage production was observed in *E. coli* K-12(λ)$^+$ phage when the flight lasted more than one day. However, a correlation between magnitude of the effect and spaceflight duration could not be established [37].

No effect of spaceflight on fibrolasts and A-1 cells was observed. Indeed, a HeLa cell strain repeatedly exposed to spaceflight conditions for six times (HeLa 19 cells) showed in space a viability higher than that of HeLa cells flown only once. The HeLa 19 strain displayed a remarkable increase of cell size which was 1.5 times higher than that of control cells. This property persisted for at least eight passages. No details are given on the conditions of incubation during flight [54].

The advent of biosatellites (Table 2) opened the way to an extensive international cooperation between the USA, France and a number of Eeast European countries. The biological material was proceeded at the recovery site in transportable laboratories set up in tents. Particularly interesting were the facilities for the recovery of the Cosmos 936 and 1129 biosatellites. The preferred test animal was the rat. Dramatic physiological changes due to weightlessness were detected. There is an overall parallelism between the effects observed in man, namely the space motion sickness, cardiovascular changes, degradation of bone material and blood changes (Sect. 2.1) and those caused in rats. In addition, changes were detected in several organs like the thymus, spleen and bones. However, when the animals are kept in a centrifuge generating artificial $1 \times g$ gravity during flight, most of the changes are eliminated or at least attenuated.

With the launch of the large-space laboratories of the type Salyut, an extensive program of biological experiments was started (Table 3). Several of those performed on Salyut-6 are still under evaluation and the results are not yet available. One of the most interesting project was a feasibility study of a fully autonomous greenhouse, with the purpose of having in space a system capable of delivering fresh and edible vegetables during long-duration flights. Several attempts have shown that higher plants do not normally develop in the weightless environment. However, when artificial $1 \times g$ gravity is provided, plants develop almost normally. This has been tested with the "Gravistat" equipment [61, 62]. As seen on previous flights, little effect of spaceflight is detected on microorganisms. An exception was given by *Paramecium aurelia* in the Cytos experiment of Salyut-6 [59, 60]: The microorganism grew faster at $0 \times g$. Practically no results are available on the growth and adaptation of mammalian cells in space. To the best of our knowledge, the only experiment with animal cells was performed with Chinese hamster cells on Salyut-6.

Operations on board of Salyut-6 are being continued in 1981 and other large stations hosting up to 12 cosmonauts will be launched by the Soviets in the future.

With regard to flight hardware, two incubator units were described in more detail than usual: the biotherm and biotherm-1 thermostats [56] which were flown on the Salyut-4 [55] and Salyut-5 [58] stations (Table 3). Both instruments are constructed on the basis of Peltier elements consisting of semiconductor thermoelectric cells as cooling/heating units. Biotherm works in a temperature range between 15 and 30 °C, has an internal volume of 100 ml, weights 9.5 kg and is alimented by batteries for up to 15 days. Biotherm-1 has a temperature range of 20—35 °C, a useful volume of 350 ml, and a weight of 2 kg. Power consumption is 3 W either from on-board 27 V DC current or from batteries.

In conclusion, one of the most important achievements of Soviet activities in space biotechnology was the finding that artificial gravity restores favorable conditions in space for the adaptation, development and survival of higher plants and animals. This is of primary importance for the establishment of large inhabited space stations in the next decades. However, artificial gravity has never been tested on humans.

5 The Spacelab

Spacelab is a manned orbital laboratory providing a pressurized module and an unpressurized platform (pallet) with a number of standard facilities. In contrast to its predecessors Skylab and Salyut, it is a resuable system which is carried in orbit and back to earth in the cargo bay of the Space Shuttle where it remains during the entire flight (Fig. 7).

The duration of a standard mission is at present 7 days. There are plans for extending the duration of one flight up to 30 days. The crew is formed by 5 to 7 astronauts,

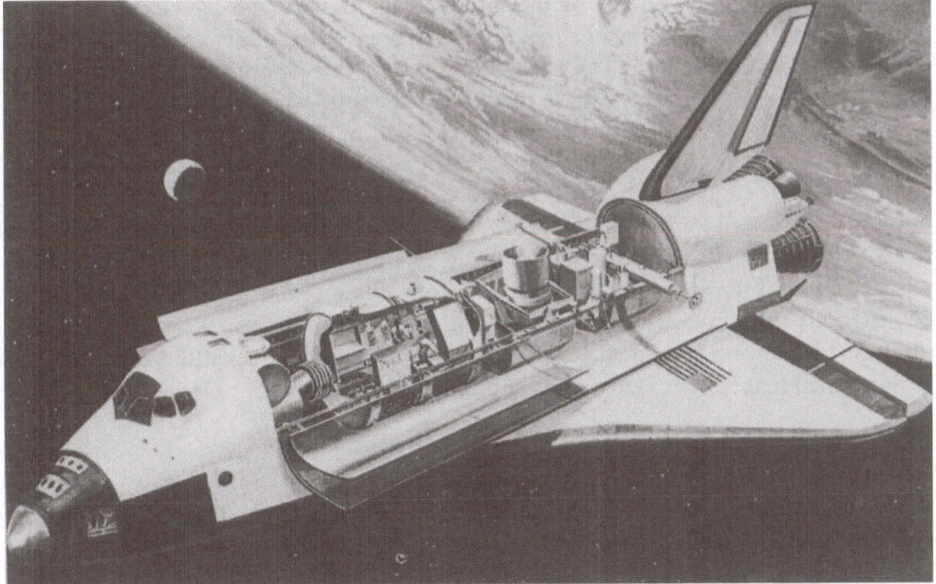

Fig. 7. Flight set-up of the Space Shuttle carrying Spacelab in orbit (from Ref. [63])

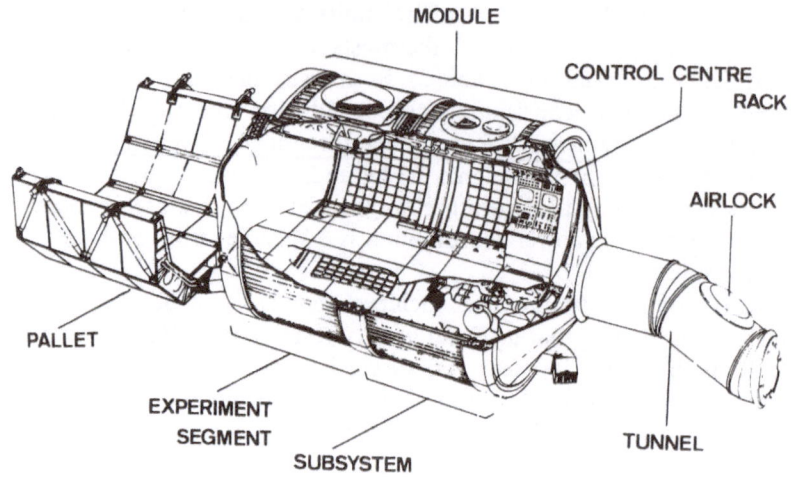

Fig. 8. Section of Spacelab in a Module/Pallet arrangement (from Ref. [62])

4 of them will operate the experiments, two as mission and two as payload specialists. Commander and pilot do not normally interfere with experiments. Instruments from different disciplines like astronomy, physics, life and material sciences, and earth observations can be accommodated in the module and on the pallet. The first mission is scheduled for 1983 and will carry 77 experiments from different disciplines. Missions devoted to life sciences are planned for the Spacelab follow-on program.

A detailed description of the Space Shuttle/Spacelab system is given in several publications of ESA and NASA (see e.g. Refs. [65–69]). Here we briefly describe the main facilities offered. The module is made of cylindrical shells and the pallet of segments (Fig. 8). The length of the module and that of the pallet can be varied according to the requirements of the mission and extended either to an "all-module" or to an "all-pallet" set-up. The hardware used for the experiments is fixed to racks in the module and on the ground of the pallet. The total weight of the payload ranges from 5.500 kg for a module/pallet to 9.100 kg for an all-pallet set-up.

The power available is either 28 V DC or 115/200 V 400 Hz, AC. The environmental control provides the module with an air temperature of 18—26 °C and a relative humidity of 30—70%. The orbit altitude ranges between 200 and 900 km. A typical flight profile of a Spacelab mission is given in Fig. 9.

A selection of the experiments is made periodically by ESA and NASA on the basis of proposals submitted by investigators in answer to announcements of flight opportunities. In addition, the agencies are ready to evaluate the scientific relevance and the technical feasibility of proposals submitted at any time.

5.1 The First Spacelab Mission

The objective of the first Spacelab Mission is to show the scientific community that the facilities offered are adequate to perform basic and applied research taking profit of the peculiar space environment. It is a multidisciplinary mission carrying experiments from Europe, USA, India and Japan. Great efforts were made by the

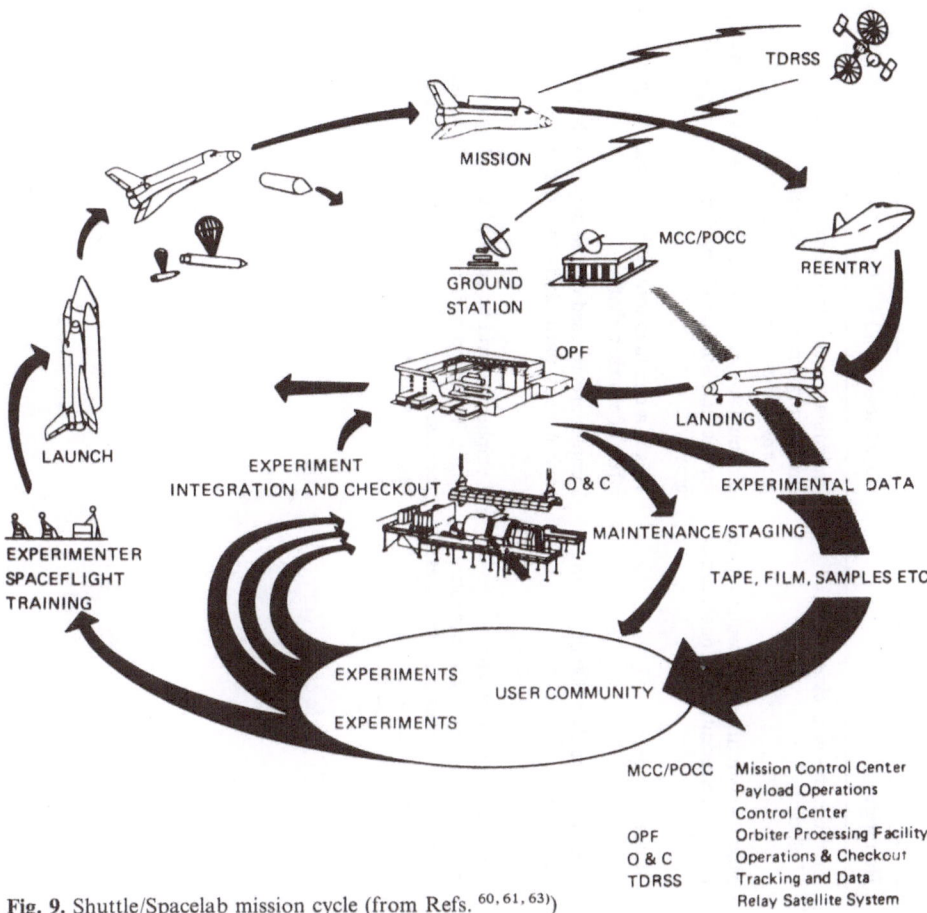

Fig. 9. Shuttle/Spacelab mission cycle (from Refs. [60, 61, 63])

MCC/POCC	Mission Control Center Payload Operations Control Center
OPF	Orbiter Processing Facility
O & C	Operations & Checkout
TDRSS	Tracking and Data Relay Satellite System

agencies and investigators to accommodate all instruments overcoming the constraints of weight, room, energy, and crew time. For the same reasons, experimental equipment and operations should be designed to minimize the use of the facilities available. Therefore, severe limits are set to the experimental resources and relatively simple experiments have a higher probability of success. Better conditions will be offered by the follow-on Spacelab program with the advent of discipline-dedicated flights.

The life science projects of Spacelab-1 include physiological investigations on humans (crew members) and biological experiments on plants, microorganisms and cells *in vitro* [70]. The vestibular system and the space motion sickness in man will be investigated in experiments 1ES201 from Germany, 1NS102 and 1NS104 from USA. Blood specimens will be withdrawn inflight and analyzed later on the ground for red cell mass reduction (1NS103, USA), hormonal levels (1ES026 and 1ES032, Germany) and immunological parameters (1NS105, USA). Periodic motions of the human body, resulting from heart activity and blood ejection into the aorta, will be recorded by three-dimension ballistocardiography (1ES028, Italy). Several physiolo-

Table 4. Biological experiments on the first spacelab mission

Experiment	Biological object	Hardware	Objectives
1NS007 (USA)	*Neurospora crassa*	Passive carry-on package of plexiglas	Preliminary characterization of preexisting circadian rythmus during spaceflight. *Neurospora crassa* as model system
1NS101 (USA)	*Helianthus annuus* seeds	Modules with dry seeds, water added inflight, $1 \times g$ centrifuge, photocamera	The nutation of sunflower is studied at $0 \times g$ and, for the first time on US payloads, at $1 \times g$ artificial gravity
1ES027 (Germany)	Several microorganisms and seeds	Advanced biostack	This is a continuation of the biostack studies on the effect of cosmic radiation
1ES029 (Germany)	Bacterial spores and vegetative cells of *Bacillus subtilis*, biomolecules like enzymes and nucleic acids	Box with more than 300 dry samples, motor, gear	Samples are exposed on the pallet to the space environment, particularly to radiation
1ES031 (Switzerland)	Cultures of human lymphocytes	Sealed culture flasks with membrane for injections, syringe rack, portable incubator (batteries and board power) at 37 °C	Lymphocytes will be exposed to mitogens during flight. The effect of $0 \times g$ on cell differentiation processes is studied by biochemical and electron microscopic methods

Fig. 10. Hardware of HEFLEX experiment for the first Spacelab mission mounted in a rack (courtesy of NASA JSC)

gical parameters will be recorded by an electrophysiological tape recorder carried by the test person throughout the mission (1ES030, United Kingdom).

The astronauts will have to discriminate between different masses in $0 \times g$ by estimating the mass of unknown samples (1ES025, United Kingdom). Dosimeters located at different sites in the module, on the pallet and in the Shuttle cabin will record the incidence of cosmic rays (1NS006, USA).

The biological experiments for the first mission are summarized in Table 4. Particularly important is the HEFLEX experiment in which, for the first time in a USA project, a $1 \times g$ centrifuge is used in space for growing plants (Fig. 10) Again, several microorganisms and seeds will be exposed to the space environment in the advance biostack experiments. The experiment with lymphocyte cultures (1ES031, Switzerland) is prepared in our laboratory and is discussed in more detail in Sect. 5.1.1.

5.1.1 Lymphocyte Cell Cultures on Spacelab

In this section we discuss the scientific background and the objectives of our investigations with lymphocytes on Spacelab. Our intent is to describe in detail a typical biological experiment aimed at providing preliminary information on an important topic related to human physiology (the immune system) and to cell biology (cell differentiation) in the weightless environment.

Post-flight biomedical studies of the crews of American and Soviet space missions have revealed diminished lymphocyte reactivity [9-14, 71]. Lymphocytes constitute approximately 30% of the white cells in human blood and play an important role in maintaining immunity against infections.

Any weakening of a space crew's immunity to infection would represent a serious hazard during and after flight and the efficiency of the immune response has therefore been tested for crew members of past missions. However, it has not yet been possible to draw a clear conclusion from the available data and further investigations are needed.

A reaction similar to that occurring *in vivo* against antigens can be triggered *in vitro* when lymphocytes are exposed to a number of substances called mitogens. Maximum activation is usually observed on the third day of cultivation. The activation can be accurately measured by incubating the cultures with tritiated thymidine. Thymidine is incorporated into activated cells at a much higher rate (100 to $200 \times$) than into resting cells.

The transition from the resting status to stimulated lymphocytes is an example of cell differentiation. The *in vitro* activation of lymphocytes by mitogens can therefore be regarded as a good model for testing the efficiency of the immune response and for the study of the mechanism of cell differentiation, the latter being one of the most interesting topics in biology today. The objective of our experiment on Spacelab-1 (experiment number 1ES031, titled "the effect of weightlessness on lymphocyte proliferation") is to test the reactivity of human lymphocytes *in vitro* during the spaceflight [72, 73].

The essential operations of the experiment are outlined in Fig. 11. Our team will be responsible for all ground operations at Kennedy Space Center, at a biological laboratory provided by NASA. On launch day, venous blood will be taken from a healthy donor. The lymphocytes will be purified by standard procedures and introduced into culture. One portion of the culture will be retained for a ground-

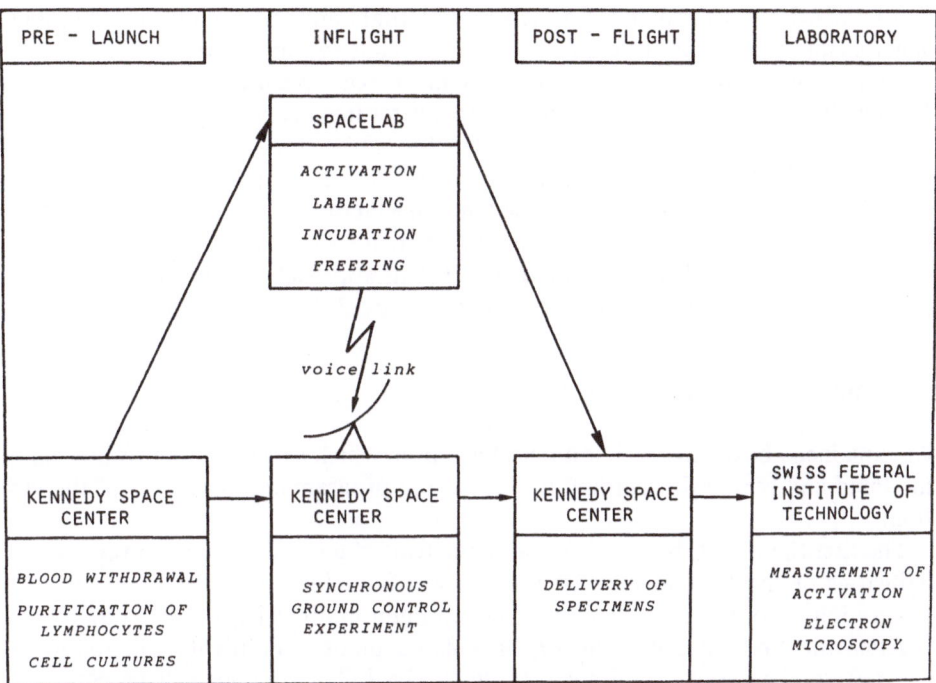

Fig. 11. The logistic of experiment 1ESO31 for the first Spacelab mission: "Effect of weightlessness on lymphocyte proliferation" (from Ref. [72])

control experiment, the other portion will be sealed in four cell culture chambers. The cultures are to be placed in a carry-on incubator kept at 37 °C which will be stowed in the Shuttle orbiter cabin 8 h before take-off. The hardware is discussed in detail in Sect. 7.

Once the Shuttle/Spacelab are in orbit, a crew member will activate the experiment by injecting the mitogen concanavalin A into the four chambers using the syringes contained in the incubator. When the Spacelab module becomes accessible, about 12 h after launch, the incubator will be transferred to the module, fixed to its front panel mounted into rack no. 4 and connected with the on-board power.

After 70 h of incubation at 37 °C during which the experiment will remain unattended, the radioactive ^{3}H-thymidine will be added to the cultures. After a further 2 h of incubation, hydroxyethylstarch will be injected into the cultures to preserve cell ultrastructure during freezing. The culture block will then be removed from the incubator and stowed in a freezer precooled by liquid nitrogen. The biological specimens will be handed out to the investigators within 24 h of the Shuttle's landing.

The parallel ground-control experiment will be run with 1 h delay, coordinated by a voice link with the Spacelab crew, to allow for possible blackouts and contingencies.

The frozen cells will be transported to our laboratory in Zürich. After careful thawing, cell activation will be determined by measuring the amount of radioactivity incorporated into each cell culture. Ultrastructure will be analyzed by thin section, freeze fracture, and scanning electron microscopy. We will also recover some viable cells and try to culture them further in vitro.

A comparison of the data from the Spacelab flight and the ground should provide valuable information on the efficiency of the immune system in space.

This study will be probably followed by more extensive investigations on further Spacelab missions. A project in which lymphocytes from crew members are analyzed pre-, in- and post-flight has been selected by NASA for a life science-dedicated mission and by ESA for a biorack mission (Sect. 5.2) as candidate for flight.

Similar experiments can be performed in space laboratories with cells of the blood system, e.g. by studying the transformation of stem cells from the bone marrow into reticulocytes and finally into erythrocytes [71]. This approach should contribute to the understanding of the hematological changes observed during spaceflight.

5.2 The Follow-on Spacelab Program

As explained above, Spacelab is a reusable unit in which experiment hardware can be mounted and removed according to the specific program and objectives of the mission.

The first flight unit, built in Europe, is the fruit of a collaboration between several industries of the ESA member states and will be delivered to NASA in 1982 for integration in to the Space Shuttle. A second flight unit has been ordered by NASA.

After the first Spacelab mission in 1983, a number of flights will follow. In contrast to the first multidisciplinary flight, the following ones will be concerned with only one or two disciplines like the NASA Life Sciences dedicated mission which should take place 12–24 months after the first Spacelab flight.

On the other hand, ESA is actually planning the manufacture of a "biorack" to be integrated in the Spacelab. Biorack is a multi-user facility capable of accommodating several biological experiments. Another multidisciplinary mission, including life sciences, will be sponsored by the German Government after Spacelab-1.

Unfortunately, the cuts in the budget of NASA, made by the new US Government, and the persistent uncertainity of ESA on the future of the Spacelab do not allow, for the time being, to make a reliable forecast on the schedule of the future Spacelab flights. This situation is also due to the delay of the first Space Shuttle test flight and the demonstration of its operational readiness.

Hopefully, after the first successful flight of the Orbiter Columbia in April 1981, financial support will be granted with more enthusiasm to the development of basic and applied research in space.

The following sections will give a description of the principal hardware presently developed by the two agencies for biological studies on Spacelab.

5.2.1 Equipment Provided by NASA

A significant effort is made by NASA to provide extended facilities for biological experiments on Spacelab. The Life Sciences Laboratory Equipment (LSLE) is described in Ref. [74]; however, engineering details and designs are missing. The typical biological equipment is summarized in Table 5. In addition, auxiliary equipment like mass determination devices, photo- and picture-cameras, video recorders and computers will be available. A number of other items for recording physiological parameters on humans and animals are included in the LSLE.

Table 5. NASA life sciences laboratory equipment for biological experiments

Item		Power	Weight (kg)
Orbiter centrifuge	$1400 \times g$, 12 tubes 15 ml each, speed not variable, for separation of blood components	115/200 V 1.3 amp, 400 HZ	11.4
Laboratory centrifuge	Two rotors: Bucket type for $50–1600 \times g$, 35°-fixed angle rotor up to $4000 \times g$, 48–10 tubes of 1–50 ml	28 V DC, 840 W	30
Hematocrit centrifuge	Operates at 0–40 °C, stowage at $-20°/+55$ °C, $5400 \times g$, capillaries 32 mm length, 0.9–60 µl	six 1.5 V batteries	0.83
Active freezer	$-15°/-30$ °C, capacity 50 l	28 V DC, 310 W	25
Passive freezer	$-195°/-185$ °C, capacity 21 l		12.5
Active freezer	-70 °C, capacity: Two chambers, ambient pressure 40 l, vacuum 25 l	28 V DC, 600 W	8.5
Refrigerator/freezer	$-22°/+10$ °C, capacity 71 l	28 V DC, 350 W	34.9
Blood-collection kit	15 and 10 ml corvac vacutainer syringes, 5 ml monovettes		8
Mass spectrometer gas analyzer	For N_2, O_2, CO_2, H_2O, A, He, C_2H_2, total hydrocarbons	28 V DC, 115 W	35
Incubator	capacity 28–42 l, from 1 °C above ambient temp. to 55 °C, culture dishes	28 V DC, 80 W	36
Centrifugal fast analyzer	Rapid analysis of biochemical samples: Variable speed rotor, temp. control, photometric system, micro-processor, printer	28 V DC, 250 W	41
General purpose working station	Working area, equipment and instruments for several biological experiments		
Research holding facility	Holding for small animals: 24 rats, or 144 mice, or 4 squirrel monkeys, $0 \times g$ or $1 \times g$ operations	320 W	200
Microscope system	Trinocular bright/dark field, phase contrast, $40–1000 \times$, trinocular stereo-dissecting microscope, zoom lens $2.5–280 \times$	28 V DC 115–200 V AC	15
Rodent sacrificing/ blood collection kit	Cervical dislocation device, decapitator, vacuum, pressure or centrifugation device for blood collection	28 V DC 115–200 V AC	15
Biological specimen test apparatus	20 and $40 \times$ time-lapse photomicro-graph, temp. 37 °C	28 V DC, 16 W	10
Slide staining system	Stainer for bacterial and blood smears		2.2

26 experiments were tentatively selected by NASA for flight on a life science mission, most of them from the USA and two from Europe. One is our proposal devoted to the study of the behavior of lymphocytes in space.

5.2.2 The ESA Biorack

The term "biorack" applies to a multiuser facility for biological investigations on Spacelab. The biorack is designed to host experiments on cell and developmental biology, radiobiology and exobiology. A first feasibility study on incubators and holding units for cells and tissues, plants and low vertebrates was performed by Dornier (Germany) and Steel (France) by order of ESA in 1977 [75]. The result of the study was a detailed description of four incubator units capable of accommodating: a) Fishes at two temperature ranges, 6–12 °C and 20–25 °C, in a useful volume of 0.13 m^3, weight 30 kg, power 30 W. A camera could record the movements of the animals. b) Frogs at 4–12 °C, in a useful volume of 4 l, power 25 W, weight 15 kg, c) Cells and tissues at 15–40 °C, volume 1.7 l, humidity 40–95 °C with up to 5% CO_2, the pH could be recorded, power 15 W. d) Plants at 5–25 °C in 50 l volume, 60–95% humidity, weight 38 kg, power 60 W. The study did not include accessory equipment for biological experiments like microscope, centrifuge, freezer, and refrigerator.

On the basis of this study, ESA issued an announcement of opportunities in 1978. 15 experiments were selected as candidates and used as a model payload for a new biorack design. ESA decided to eliminate the plant, fish and frog holding units, in favor of two incubators for cell cultures and microorganisms, microscope, and centrifuge [76].

The incubators have temperatures ranging from 2 to 27 °C and from 27 to 37 °C, respectively. Control of CO_2 partial pressure and humidity is ensured. The useful volume per unit is 2 l. The biorack is completed by a freezer 0 to −25 °C, a high-power microscope (1250 X), a low-power stereomicroscope (400 X), a centrifuge above 200 × g and a 1 × g reference centrifuge. However, the biorack is not defined in its last details and further inputs from the investigators are required. A new call for proposals was issued by ESA in 1980. A number of experiments were selected according to categories of priority. It will be on the basis of the final selection that the final experimental set-up of the biorack will be defined. A decision on funding will be taken by the ESA member states in 1981/82.

5.2.3 Other Biological Equipment for the Spacelab

Several laboratories and industries are presently developing flight hardware for biological projects to be installed on Spacelab or on the Space Shuttle (in the flight middeck). The advantage of mounting hardware in the cabin of the Shuttle is that the experiment can be performed more frequently, without waiting for a Spacelab mission. However, the room available in the Shuttle cabin is very limited.

MBB in Germany is developing a new model of the continuous flow electrophoresis equipment which was successfully tested on the ASTP mission (see Sect. 3.3.2). The technical characteristics are: Flow rate of the sample: 1–10 ml h^{-1}; electrical potential: 20–80 V cm^{-1}; cooling of the separating chamber: 5 °C; maximum sample deviation: 30 mm; resolution: 0.1 mm analytical, 0.5 mm preparative.

Another project envisages the processing of human and animal collagen fibers for the manufacture of protheses. The pharmaceutical industries are particularly interested in the application of electrophoresis to the separation of urokinase-producing cells (see also Sect. 3.3.1 and 6).

The National Space Development Agency of Japan (NASDA) is presently investigating the possibility of launching biological payloads in space, either with a vector developed in Japan or in collaboration with ESA and NASA. In conclusion, we can forecast that a successful completion of the Shuttle test flights and of the first Spacelab mission will give great impulses to the development of equipment and techniques for biological experiments and applications in space.

6 Bioprocessing of Mammalian Cells in Space

One of the most ambitious programs of NASA in the field of space biotechnology is devoted to the processing of animal cells in space. The instrumentation is designed to be installed either in the Shuttle cabin or in the Spacelab module. It consists essentially of two subsystems. One is a bioreactor for growing cells at $0 \times g$ and the other a continuous-flow electrophoresis apparatus for the separation of cells. A block diagram of the bioprocessing unit is shown in Figs. 12 and 13. The electrophoretic system takes advantage of better separation conditions in the weightless environment as discussed in Sect. 3.3 and is a further development of the equipment successfully tested on the Apollo 14 and 16 and on the ASTP missions. The new electrophoresis unit will be tested on one of the Space Shuttle test flights.

While the advantage of separating cells by electrophoresis in space has been well demonstrated, there are not yet many arguments in favor of $0 \times g$ bioreactors. Nevertheless, we can make some interesting considerations. The lack of sedimentations permits a homogeneous distribution of the cells in the culture. Consequently, the system can work with a higher cell density and a better exploitation of the medium than at $1 \times g$. In addition, the unexperienced weightless environment can influence cell growth and activity with the potential benefit of delivering new useful products or higher yields of known substances. In this context, we have seen that gravity may interfere with the cell cycle of human lymphocytes and of transformed cells (see Sect. 8).

A prototype of a bioreactor for Spacelab is presently tested in the biological laboratories of JSC (D. Morrison, personal communication). The test unit is of the type used in most laboratories. The reactor consists of a jacketed glass vessel connected to a water heater controlling the temperature in a range of 30–40 °C. The first problem is the choice of a suitable material for the container since glass is undesired in space laboratories. Secondly, the absence of gravity implies lack of convection and this may create serious temperature control problems even in pumped heating systems.

Within the bioreactor the media are mixed by means of a spin filter. This also serves to filter out the media from cells or microcarriers. The media continuously circulate through a media-reconditioning loop where they are pumped countercurrently to fresh media through dialysis tubes. Reconditioned media flow back into the reservoir where redox potential, dissolved oxygen, pH, and temperature are measured. The respective values are sent through Burr-Brown amplifiers to the A/D board of a LSI-11 microcomputer and stored on floppy.

The stirring rate of the spin filter is very critical and related to well-known problems like clumping of the beads and high shear rates damaging beads and

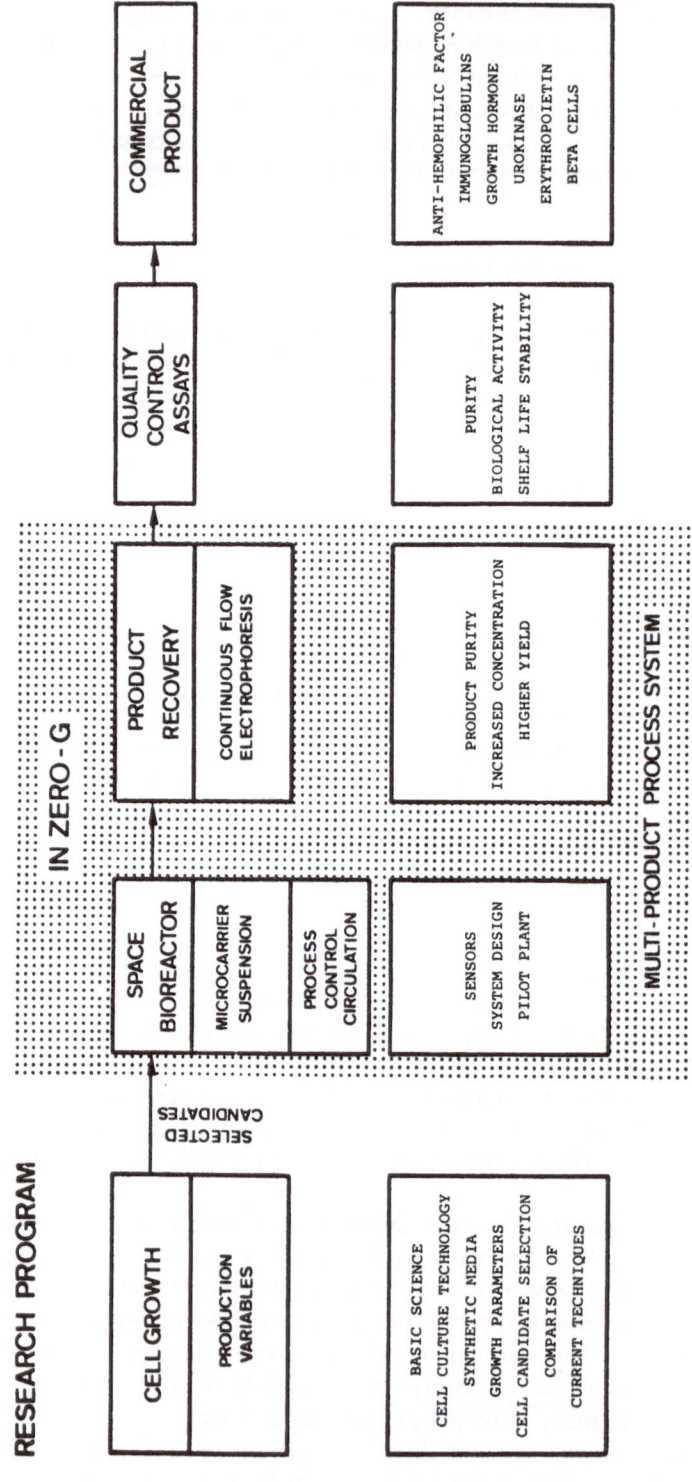

Fig. 12. Conceptual block diagram of biosynthesis in space (courtesy of NASA JSC)

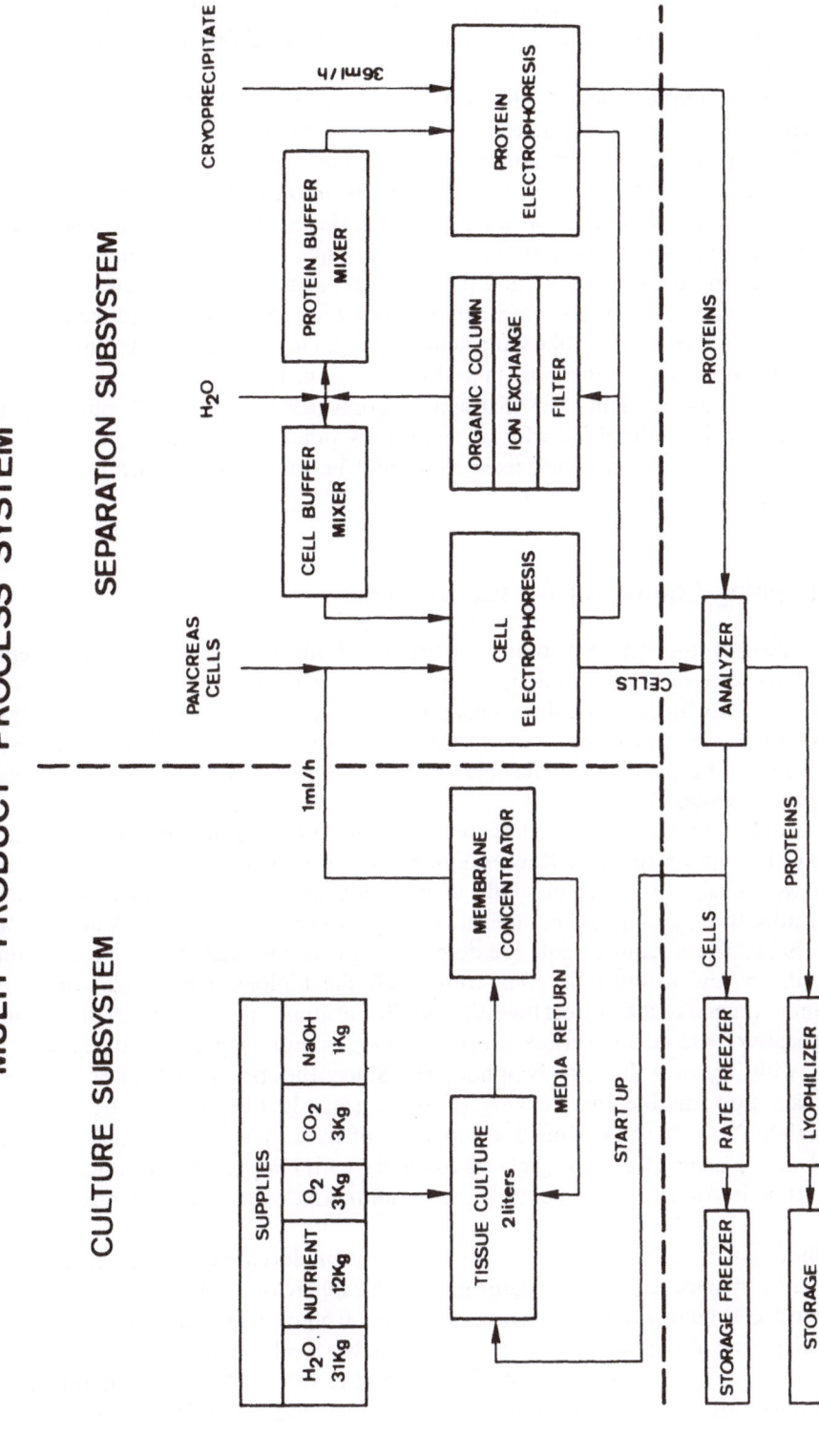

Fig. 13. Detailed diagram of the multiproduct process system shown in Fig. 13 (courtesy of NASA JSC)

cells. However, a $0 \times g$ environment may require less vigorous stirring and thus contribute to solve one of the most critical problems of culturing mammalian cells in bioreactors.

Technological applications in this field should be preceded by a program of basic research on the behavior of interesting "candidate cell systems" in a space environment. Preliminary studies in incubators like those provided by biorack or the LSLE are necessary. In addition, useful information on gravitational effects can be gained by ground simulations at high g in centrifuges and low g in clinostats (see Sect. 8). Our experiments on Spacelab and under altered g conditions with lymphocytes and other cell lines in vitro are a first step in this direction.

Potential candidates for bioprocessing in space are cells secreting products of high pharmacological relevance like anti-hemophilic factor, immunoglobulins, growth hormones, urokinase, erythropoietin and insulin (Fig. 12).

We are well aware of the fact that similar goals can be achieved through genetic engineering, a biotechnology with an enormous potential development. However, at present, a forecast on which technology will be the more successfull is not yet possible.

7 Designing Equipment for the Spacelab

The purpose of this section is to provide first-hand information on the requirements, constraints, interfaces involved in the preparation of experiments, and hardware for the Spacelab. Since we are directing and managing several projects either finally or preliminarily selected for flight, we discuss this topic on the basis of the experiences gained during the last 4 years. The objectives and scientific background of our experiments are described in Sect. 5.1.1.

ESA and NASA set strict safety requirements on materials, electric parts and set-up of the equipment to be flown on manned space mission. Although our experiments were designed to require only a minimum of crew operations, weight and energy allocation, a long and complicated integration procedure by ESA and NASA is necessary. These requirements are described in detail in Refs. [77, 78]. On the other hand, the hardware must be compatible with the biological demands which are sometimes crucial of the objects investigated. For example, flasks for cell culture must have peculiar material and surface properties. One example is provided by our experiments with *in vitro* human lymphocytes. Since the traditional plastic culture flasks are not suitable for working in space (mainly due to their fragility and flammability), we have developed a new kind of flask. We tested several materials (metals and plastics) for biological compatibility with lymphocytes in order to select those suitable for our equipment. Table 6 summarizes the results of our material tests.

Another problem was raised by the necessity of developing a culture medium autonomously buffered, i.e. not requiring an atmosphere of 5–10% CO_2 as is the case with most cell culture media, since a device-controlling CO_2 pressure would have raised the costs of the hardware above a reasonable level. We have found a medium which provides optimum conditions for the culture of lymphocytes in hermetically sealed flasks: RPMI 1640 medium, pH 7.3, supplemented with 20% human serum,

Table 6. Materials tested for biocompatibility with lymphocytes[a]

Material	Dimension		Activation %	
	Shape	(mm)	Human	Mouse
None (control)			100	100
Aluminium anticorodal	Prism	$10 \times 12 \times 10$	81	100
Aluminium anticorodal anodized	Plate	$41 \times 11 \times 1$	58	98
Steel	Screw	$\varnothing = 5$ $1 = 11$	80	82
Rust-resistant steel	Bar	$\varnothing = 11$ $1 = 15$	56	62
Spring steel	Spring	$\varnothing = 1$ $1 = 103$	9	0
Stainless steel	Bar	$\varnothing = 15$ $1 = 50$	78	—
Spring steel optaloy	Spring	$\varnothing = 1$ $1 = 157$	17	24
Argentine	Valve	$\varnothing = 4$ $1 = 20$	71	52
Titan	Plate	$5 \times 63 \times 1$	85	45
Cadmium-coated steel (10 µm)	Washer	$\varnothing = 8$	24	100
Silver-coated steel (0.5 µm)	Screw	$\varnothing = 4$ $1 = 12$	11	10
Gold-coated steel (0.5 µm)	Screw	$\varnothing = 4$ $1 = 12$	18	6
Gold-coated steel (10 µm)	Coil	$\varnothing = 1$ $1 = 66$	95	100
Gold/Rhodium-coated steel (0.5 µm)	Coil	$\varnothing = 1$ $1 = 66$	97	100
Silicon (Silicoset caoutchouc ICI 105)	Piece	0.3 g	2	0
Silicon Lute (SIKA, Zürich)	Piece	1 g	2	0
Silicon Lute solid (Adhesa)	Piece	0.5 g	35	—
Silicon Lute liquid (Adhesa)	Clot	0.5 g	—	0
Silicon Red	Ring	$\varnothing = 26$	84	80
Silicon Blue	Ring	$\varnothing = 26$	74	100
Silicon Transparent	Tube	$\varnothing = 4$ $1 = 30$	79	70
Viton M 71	Ring	$\varnothing = 45$	83	59
EPT 77	Ring	$\varnothing = 40$	2	0
Latex (from *Hevea brasiliensis*)	Ring	$\varnothing = 40$	25	39
PVC	Tube	$\varnothing = 3$ $1 = 42$	2	0
BIO-Folie Petriperm	Piece	0.1 g	77	77
Araldit (Epoxiresin)	Piece	0.4 g	21	0
Teflon (PTFE)	Plate	$5 \times 30 \times 1$	62	81
Teflon	Bar	$\varnothing = 5$ $1 = 40$	89	100
Teflon	Ribbon	12×30	72	—
Teflon/Glass fiber 25%	Prism	$20 \times 10 \times 10$	100	—
Polyacetal Delrin	Bar	$\varnothing = 5$ $1 = 30$	83	69
Halar PETFE	Screw	$\varnothing = 5$ $1 = 5$	59	100[b]
Garlock Lute Solid	Piece	1 g	—	66[b]
Garlock-coated Halar (1 mm)	Screw	$\varnothing = 5$ $1 = 5$	48	82[b]
Garlock liquid	Clot	1 g	—	0[b]

[a] In the test incubating lymphocytes were used in the presence of the material with the mitogen concanavalin A for 3 days at 37 °C. Activation was measured by incorporation of labeled thymidine into DNA

[b] Tested on rat lymphocytes

buffered with 40 mM Hepes and 5 mM HCO_3 and containing 50 µg/ml gentamycin. This approach, i.e. the search for buffer systems not requiring gas control and maintaining a given pH level for several days, could save money and increase the chances of a suitable selection of an experiment. The culture flasks of our experiment are made of Teflon reinforced with 25% glass fiber (Figs. 14, 15). Basically, they consist of a cylindrical container which can be filled with maximally 12 ml of

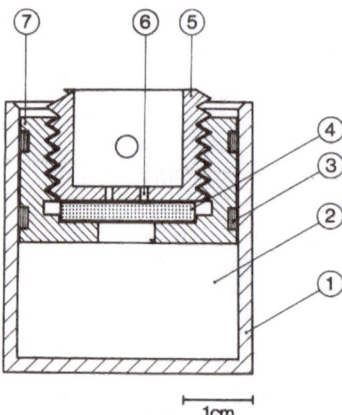

Fig. 14. Section of a cell culture flask of experiment 1ESO31. 1 Container, 2 culture, 3 sealing ring, 4 silicone rubber membrane, 5 fitting, 6 holes for injections, 7 piston

Fig. 15. Flight hardware for experiment 1ESO31. Incubator with cell-culture flask and culture block

culture and hermetically sealed with a piston. Reagents can be injected into the chamber by means of syringes through a thick membrane made of silicone rubber. The piston can move up and down, thus compensating variations of volume. The culture chambers are fixed in a culture block of aluminum (Fig. 16). The block is kept in a portable incubator. This incubator (Figs. 15, 16) consists of a carry-on box

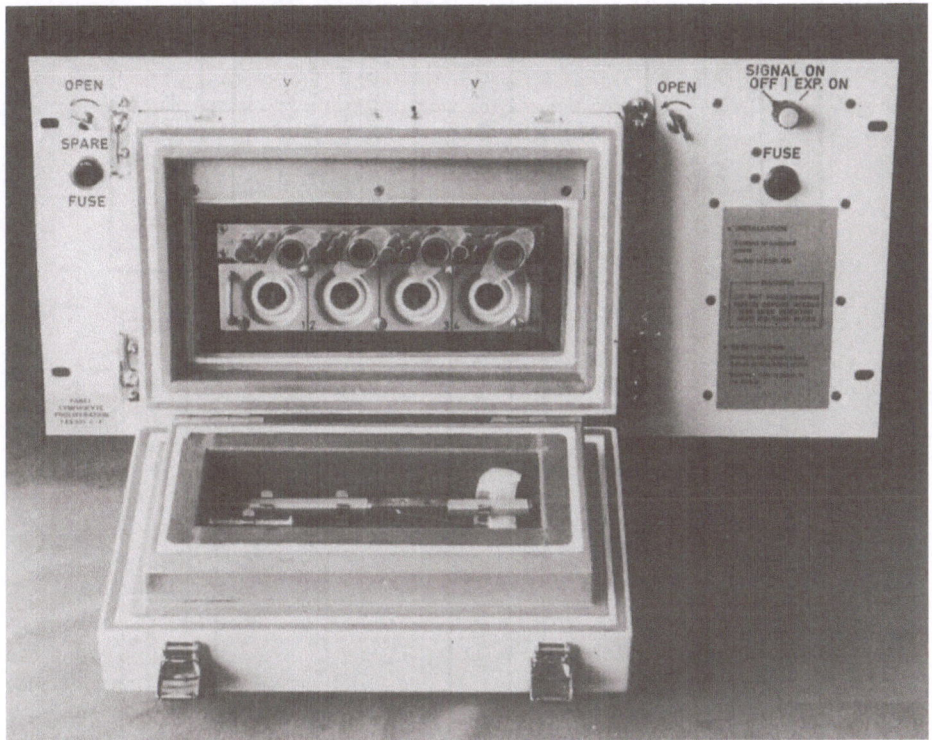

Fig. 16. Flight hardware for experiment 1ESO31. Open incubator fixed in front panel

$(25 \times 17 \times 17 \text{ cm}^3$ and weighing 3.5 kg) in which the temperature can be kept at 37 °C, either by means of a battery power (up to 24 h) or a Spacelab power (28V DC). The power consumption is 3.5 W. The incubator can be fixed to its front panel (Fig. 16) mounted in a rack in the Spacelab module. The front panel carries the electronic box with the connectors to Spacelab's power bus and to the remote acquisition unit (RAU). The RAU connection delivers a temperature signal to the ground control station at Johnson Space Center.

ESA and NASA accept hardware for flight only after passing a number of space qualification tests.

Many tests can be performed in qualified laboratories of space-oriented industries or at ESTEC (European Space Technology Center [79)]. For example, our equipment had to pass vibration, EMC (electromagnetic contamination), ITE (interface test equipment), and off-gassing tests. Safety requirements, test procedures, limits of weight, and energy consumption have a significant impact on the costs of the flight hardware. Generally speaking, every instrument must be designed and developed as it were a completely new item. As a rule, one can multiply the costs of an instrument routinely available for the ground laboratory by a factor of 10 to 50. Therefore, it will be of primary importance for the future of space science that ESA and NASA provide the investigators on the Spacelab with common multiuser facilities. Another important aspect of space-biological activities is the organization of pre-, in-, and

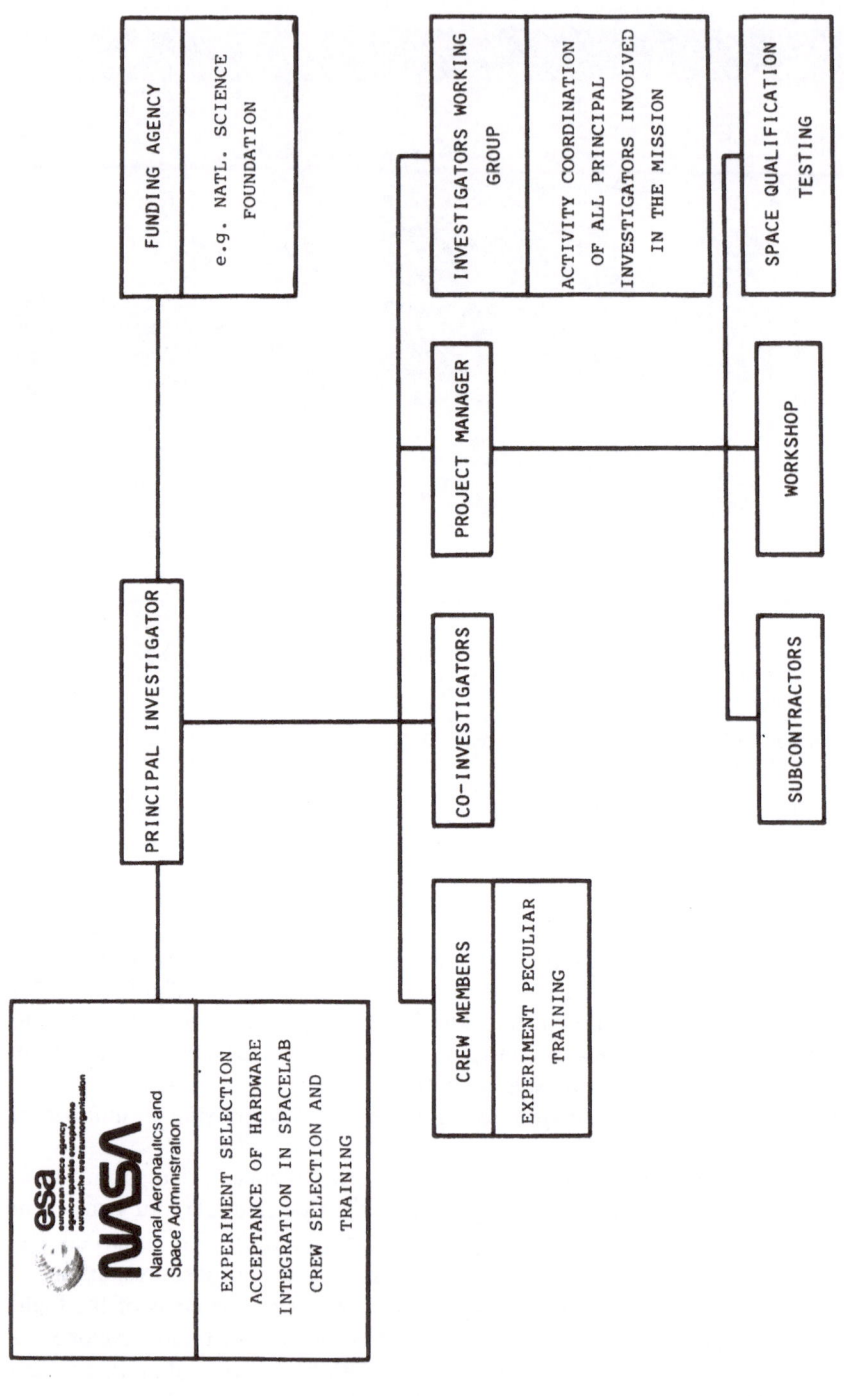

Fig. 17. Typical management structure of a project for Spacelab

postflight operations at the launch site. While most of the investigators of other disciplines are directing the operations of their experiments from the POCC (Payload Operations and Control Center) in Johnson Space Center (JSC), several life scientists need laboratory facilities at the launch site in Cape Canaveral for the preparation of biological specimens. NASA will provide standard biolaboratory facilities if these have been negotiated in time with the agency. The flight operations are coordinated by JSC in Houston. In the POCC, consoles permitting a direct voice link with the crew are available to the investigators.

One of the most critical point of biological experiments is the necessity of carrying on board the specimens as late as possible before launch (late access). Since the Spacelab is closed in the cargo bay of the Space Shuttle standing on the launch pad, no access to the Spacelab module is possible during the last seven weeks before the start of the mission. NASA responsibles are rather reluctant to concede intereferences (storage of experiment hardware) with launch operations in the last hours. Therefore, late access must be negotiated with NASA. The incubator with our lymphocyte culture will be stowed in the flight mid-deck of the Shuttle cabin 8 h before start. A similar agreèment must be bargained for early access (4–12 h) to biological samples immediately after landing of the spacecraft. In our opinion, these difficulties are mainly due to the fact that biology is a new discipline in space science and NASA responsibles are not yet familiar with late and early access requirements. Probably, after the first Spacelab flight, these requirements will be routinely integrated in launch and landing operations. The investigators are also responsible for training the crew during experimental operations: Several theoretical and practical briefings are necessary for motivating and instructing the astronauts on the scientific background and relevance of the experiment. One must take into account that working under extreme stress conditions with an overwhelming program of activities significantly raises the probability of false manipulations which could kill the experiment.

Finally, each project on Spacelab must be integrated in a complex frame of interfaces (electric, computer, mechanical, crew operations) and facilities which are used by several experiments. Hence, a complicated mission can be accomplished only on the basis of a cooperative spirit (Fig. 17).

8 High-g and Low-g Ground Simulations

Very little is known on the effect of the variation of the gravitational environment on the cell cycle, growth and differentiation of animal cells. As mentioned above, only one experiment with mammalian cells in space has been described in detail (Sect. 3.2).

The design and planning of biological projects in space should be accompanied, whenever possible, by extensive simulation experiments on the ground.

We used this approach in the study of lymphocyte differentiation during exposure of human lymphocytes to mitogens [80–84]. High g's were simulated in a centrifuge kept in a breeding room at 37 °C. A sophisticated system for high g simulations has been described by Smith et al. [85]. It consists essentially of a rotating arm providing the required g level and carrying a container for biological samples like cells, microorganisms or small animals. The container itself rotates in a direction perpendicular to that of the centrifugal force by means of a small motor. This

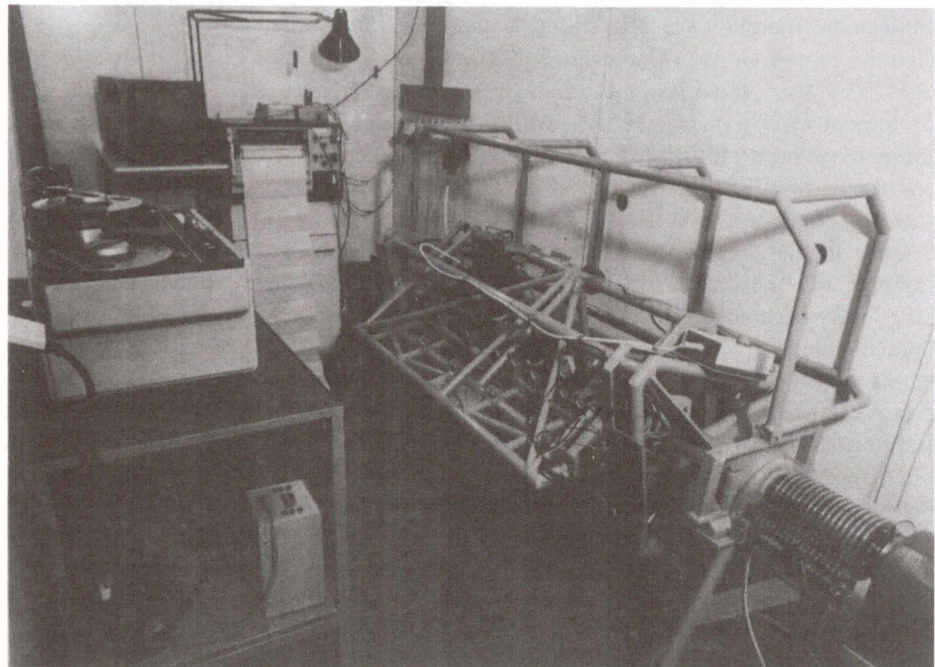

Fig. 18. Fast rotating clinostat for low-g simulations on cells and microorganisms. Right: Rotating unit with microscope carrying culture chamber, TV camera and time-lapse picture camera. Left: TV monitor and time-lapse video recorder (courtesy of W. Briegleb, DFVLR, Cologne)

avoids artefacts due to the pressure of the specimens against the wall of the container. At high g, lymphocyte differentiation is accelerated: Maximum activation occurs on the second day of cultivation instead of the third day as usually observed [80-82]. This effect is much more dramatic in transformed SGS-3 cells (a cell line from a sarcoma growing on a variety of *Rattus norvegicus* rats, namely Galliera rats). At 4 and $20 \times g$, cell growth is twice as high that of $1 \times g$ controls (unpublished observations).

A low-g environment cannot be reproduced on earth. Nevertheless, gravity can be compensated in the fast rotating clinostat [81, 86, 87]. Small animals, microorganisms and cells can be grown in small cylindrical chambers rotating along their longitudinal axis at 90–100 rpm. Gravity is compensated by rotation, provided that the diameter of the tube is smaller than 5 mm.

A sophisticated clinostat has been built by Briegleb (Fig. 18) and consists of a rotating microscope connected to a TV and a time-lapse picture camera. Our experiments with lymphocytes in the clinostat show that activation is stopped at an early stage in approximately 50% of the cells Fig. 19 [71, 81-84]. No effect of low-g has been detected in SGS-3 cells.

A further application of this approach could be the study of combined gravity and radiation effects. A synergism between weightlessness and radiation has been observed on organisms flown on biosatellites (Table 2).

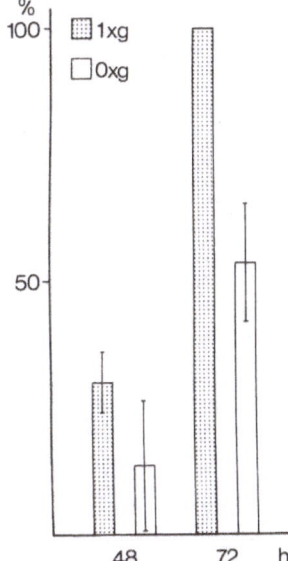

Fig. 19. Effect of simulated $0 \times g$ on lymphocytic activation. Human lymphocytes were exposed to mitogenic concentrations of concanavalin A. Activation was measured after 2 and 3 days of incubation by incorporation of labeled thymidine into DNA. $0 \times g$ conditions were simulated in a fast rotating clinostat.
The results are expressed as percent of activation, taking as 100% activation that of the control at $1 \times g$ on the third day of cultivation. The standard deviations are calculated from 4 experiments with lymphocytes from different donors

We have extended our investigations on the effect of physical stress on cells by exposing lymphocytes to radiation and vibration. By stress we mean the unusual deviation of one or more parameters which define the environment of a living system from physiological conditions.

When lymphocytes are cultivated with concanavalin A in a flask subjected to vibrations at 500 Hz, the activation is remarkably higher than in the control test. This too could be due to a g-effect generated by the vibration. Conversely, when cultures are kept under pressure (3 or 10×10^5 Pa), a depressing effect is observed, indicating that the high-g effect is not due to an increase of the hydrostatic pressure in the culture.

Irradiation with X-rays (50 rad, 200 kV) reduces activation by 30%. When the combined effects of radiation and hypogravity or of radiation and hypergravity are investigated on lymphocytes, no synergism is observed [82, 83]

Table 7 compiles the qualitative results of these experiments.

The preliminary findings that high g can enhance cell growth may have an important impact on biotechnological applications of mammalian cell cultures. Investigations in this direction are progressing in our laboratory.

Finally, the gravitational effects on the hemopoietic system and on the bones could be investigated by in-vitro simulation experiments. Important information could be gained before performing expensive and complicated experiments in space.

9 Outlook

The successful flight of the Space Shuttle Columbia has shown that man will be capable in the near future to exploit the resources available in space at a reasonable price. This will undoubtly open new perspectives to space technology. One of the

Table 7. Effect of physical stress on lymphocyte activation[a]

Conditions	Depression	Enhancement
High-g: 2, 4, 8 × g	—	+
Simulated low-g: Clinostat	+	—
Vibration: 500 Hz	—	+
Pressure: 3, 10 × 10⁵ Pa	+	—
Radiation: X-rays, 50 rad	+	—
Radiation + high-g[b]	+	—
Radiation + low-g[b]	+	—

[a] This table summarizes the qualitative results of our experiments with lymphocytes exposed to mitogens after 3 days of cultivation under conditions of physical stress: High-g, low-g, vibration, and high pressure conditions were maintained throughout incubation; irradiation of samples was performed immediately before incubation.

[b] No synergism between radiation and altered g conditions was observed

objectives of the first Spacelab flight is to show that biotechnology will considerably benefit from the facilities offered by the new space laboratories available in the next two decades.

A forecast of space technology for the period 1980–2000 was published by NASA in 1976 [88]: One of the major topics of space biology will be bioprocessing. Here, the lack of a systematic investigation on the behavior of several cell lines of interest at $0 \times g$ makes the choice of suitable biological objects rather difficult. This can be achieved only by performing preliminary studies on selected candidates and by developing an adequate laboratory hardware. The ESA biorack and the NASA LSLE are only the beginning of this technology. Experiments with cell cultures for the Spacelab have shown that a considerable amount of work has to be dedicated to the selection of biocompatible materials satisfying the safety requirements of manned spaceships. In addition, only simple manipulations are feasible in orbit and there are significant limits on weight, power and crew time.

A further step will be the foundation of space colonies on the planets and on orbital stations. This is now far from being only a subject of science fiction.

The medical aspects of man in space have been studied for 20 years. The next Spacelab flights will probably contribute to solve some of the main problems identified.

Another important topic, involving biotechnology, is the establishment of independent ecosystems, capable of recycling autonomously the resources available. The NASA-study [88] forecasts that by the year 2000 it will be possible to establish a stable, closed ecosystem with only 10–20 living species, provided that a consistent research program is undertaken in this area. About 40 plant species suitable for growth in space will be available by the same time. The farm area required per person will be reduced to 0.4 hectars. However, according to the NASA forecast, 10 years of research are necessary.

Several studies are concerned with the development of closed ecosystems for generative life support (see e.g. Refs. [89–92]). A consistent biotechnological effort is required for the development of optimum conditions of growth and the development of living systems in space.

A remarkable achievement was the establisment of a greenhouse kept under artificial $1 \times g$ conditions and supplying the crews of the Salyut-6 station with fresh vegetables. Another important step forward was the discovery that artificial gravity significantly promotes the adaptation of animals (rats) to weightlessness. Other problems of human communities in space have to be solved by sociologists, politicians and lawyers [93].

Although the technology for building large inhabited space stations has been developed, systematic engineering studies have revealed that serious failures can frequently result from a complicated construction like a space habitat.

Since even a relatively simple space mission like a Spacelab flight requires long-term planning and coordination, only a great multinational and multidisciplinary effort can guarantee an optimum and fruitful exploitation of the space resources.

We are convinced that in spite of political, technological and economic difficulties, it will be possible to realize at least part of the projects discussed in this review.

10 Acknowledgements

We wish to acknowledge the financial support of the Swiss National Science Foundation, Berne, Grant No. 3.449-0-79.

We thank H. Bücker, Frankfurt, D. R. Morrison and R. G. Thirolf, Houston, for providing us with photographic material and P. Bislin and U. Zilian for their help in the preparation of this manuscript.

11 Abbreviations and Acronyms

ASTP	Apollo-Soyuz Test Project
COSPAR	Committee on Space Research
DFVLR	Deutsche Forschungs- und Versuchsanstalt für Luft- und Raumfahrt e.V.
EMC	Electromagnetic Contamination
ESA	European Space Agency
ESTEC	European Space Research and Technology Center
HEPES	[4-(2-Hydroxyethyl)-1-Piperazineethanesulfonic Acid]
HZE	High Charge Z and High Energy
ITE	Interface Test Equipment
JSC	Johnson Space Center
KSC	Kennedy Space Center
LSLE	Life Science Laboratory Equipment
MBB	Messerschmitt-Bölkow-Blohm
NASA	National Aeronautics and Space Administration
NASDA	National Space and Development Agency of Japan
Pa	Pascal, pressure unit (1 atm = 101 325 Pa)
PI	Principal Investigator
POCC	Payload Operations Control Center

RAU   Remote Aquisition Unit
RBC   Red Blood Cells
RPMI   Roswell Park Memorial Institute
sr   Steradian (SI solid angle unit)

12 References

1. Bjursted, H.: Biology and Medicine in Space: ESA BR-01 (1979)
2. Proceedings of the Symposium Basic Problems of Man in Space, Acta Astronautica (in print)
3. Johnson, P. C., Kimzey, S. L., Driscoll, T. B.: Acta Astronaut. *2*, 311 (1975)
4. Johnson, P. C., Driscoll, T. B., Le Blanc, A. D.: NASA SP-377, 235 (1977)
5. Kimzey, S. L., Johnson, P. C.: NASA SP-411, 101 (1977)
6. Balakhovskiy, I. S., et al.: Space Biol. Aerosp. Med. *12* (3), 11 (1978)
7. Rudniy, N. M., et al.: Space Biol. Aerosp. Med. *11* (5), 37 (1977)
8. Legen'kov, V. I., et al.: Space Biol. Aerosp. Med. *11* (6), 1 (1977)
9. Yegorov, A. D.: NASA TM-76104, 173 (1979)
10. Konstantinova, I. V., et al.: Space Biol. Aerosp. Med. *7* (6), 48 (1973)
11. Soffen, G. A.: USSR Space Life Sciences Digest. Annual Summary. NASA Contract NASW-3223 (1979)
12. Kimzey, S. L., in: Biomedical Results from Skylab. Johnston, R. S., Dietlein, L. F. (eds.), NASA SP-377, p. 249 (1977)
13. Konstantinova, I. V., et al.: Space Biol. Aerosp. Med. *12* (2), 16 (1978)
14. Criswell, S. B., in: Apollo-Soyuz Test Project — Summary Science Report. NASA SP-412, p. 257 (1977)
15. Montgomery, P. O'B., et al.: In Vitro *14*, 165 (1978)
16. Pollard, E. C., in: Gravity and the Organism. Gordon, A., Cohen, M. J. (eds.), p. 25. The University of Chicago Press 1971
17. Kessler, J. O.: The Physiologist *22*, S-47 (1979)
18. Folkman, J., Moscona, A.: Nature *273*, 345 (1978)
19. Todd, P.: NASA TM-X-58191, 103 (1977)
20. Anderson, M., Rummel, J. A., Deutsch, S.: NASA TM-58217 (1979)
21. Eugster, J.: Bild Wiss. *2*, 26 (1964)
22. Bücker, H., in: Life Sciences Research in Space. ESA SP-130, p. 263. Paris 1977
23. Montgomery, P. O'B., et al., in: Biomedical Results from Skylab. Johnston, R. S., Dietlein, L. F. (eds.), NASA SP-377, p. 221 (1977)
24. Thirolf, R. G.: NASA TM X-58164 (1975)
25. Mc Kannan, E. C., et al.: NASA TM X-64611 (1971).
26. Allen, R. E., et al.: NASA TM X-73360 (1977)
27. Allen, R. E., et al., in: Apollo-Soyuz Test Project — Preliminary Science Report. NASA SP-412, p. 307 (1977)
28. Hannig, K. H., Wirth, H., Schoen, E.: NASA TM X-73360 (1977)
29. Hannig, K., Wirth, H.: Progr. Astronaut. Aeronaut. *52*, 411 (1977)
30. Viking 1. Early results. NASA SP-408, 59 (1976)
31. Horowitz, N. H., et al.: Science *194*, 1321 (1976)
32. Levin, G. V., Straat, P. A.: Science *194*, 1322 (1976)
33. Klein, H. P., et al.: Science *194*, 72 (1976)
34. Mazur, P., et al.: Space Sci. Rev. *22*, 3 (1978)
35. Oyama, V. I., et al.: Life Sci. Space Res. *16*, 3 (1978)
36. Buderer, M. D.: Russian Biospex. NASA JSC-17072 (1981)
37. Lukin, A. A., Parfenov, G. P.: Space Biol. Aerosp. Med. *2*, 1 (1973)
38. Federova, N. L.: Environ. Space Sci. *1* (3), 172 (1967)
39. Parin, V. V., et al.: Environment. Space Sci. *2* (2), 7 (1968)
40. Nikitina, I. V., Gertsuskii, D. F., Petrenko, L. M.: Environ. Space Sci. *2* (4), 264 (1968)
41. Kalandarova, M. P., et al.: Life Sci. Space Res. *14*, 179 (1976)

42. Gazenko, O. G., et al.: The Physiologist 22, S-11 (1980)
43. Vaulina, E. N., Kostina, L. N., Mashinsky, A. L.: Life Sci. Space Res. 14, 201 (1976)
44. Grigoriev, Yu. G., et al.: Life Sci. Space Res. 14, 173 (1976)
45. Zharikova, G. G., Rubin, A. B., Nemchinov, A. V.: Life Sci. Space Res. 15, 291 (1977)
46. Gurovsky, N. N., et al.: Acta Astronaut. 7, 113 (1980)
47. Rosenzweig, S. N., Souza, K. A.: Final Reports of US Experiments flown on the Soviet Satellite Cosmos 782, NASA TM-78525 (1978)
48. Grigoriev, Yu. G., et al.: Life Sci. Space Res. 16, 137 (1978)
49. Blanquet, Y., et al.: Life Sci. Space Res. 15, 165 (1977)
50. Rosenzweig, S. N., Souza, K. N.: Final Reports of US Experiments flown on the Soviet Satellite Cosmos 936, NASA TM-78526 (1978)
51. Timofeyev-Resolskiy, N. V., et al.: NASA TM-75582 (1978)
52. Parfyonov, G. P., et al.: Life Sci. Space Res. 17, 297 (1979)
53. Minutes of the US/USSR Joint Working Group on Space Biology and Medicine, 11th Meeting, Moscow, October 20—31 (1980)
54. Zhukov-Verezhnikov, N. N., et al.: Life Sci. Space Res. 9, 99 (1971)
55. Dubinin, N. P., et al.: Life Sci. Space Res. 15, 267 (1977)
56. Stil'bans, L. S., et al.: Space Biol. Aerosp. Med. 8 (6), 104 (1974)
57. Dubinin, N. P., et al.: Life Sci. Space Res. 15, 113 (1977)
58. Vaulina, E. N., et al.: Life Sci. Space Res. 17, 241 (1979)
59. Tixador, R., et al.: C.R. Acad. Sci. Paris 287, 828 (1978)
60. Planel, H., et al.: Life Sci. Space Res. 17, 139 (1979)
61. USSR Report, Space No. 2, JPRS 75111, 12 (1980)
62. USSR Report, Space No. 6, JPRS 75929, 12 (1980)
63. East Europe Report, Scientific Affairs No. 676, JPRS 75894, 1 (1980)
64. Kamin, A.: Translation on USSR Science and Technology. Physical Sciences and Technology No. 49, JPRS-71896, 26 (1978)
65. Spacelab users' guide. ESA SP-1001 (1976)
66. Seibert, G.: Life Sci. Space Res. 14, 153 (1976)
67. Spacelab. Bundesministerium für Forschung und Technologie. Bonn 1975
68. Space — Part of Europe's Environment. ESA, Paris 1977
69. Spacelab — Users Manual. ESA DP/ST(79) 3 (1979)
70. Craven, P. D. (ed.): Spacelab Mission 1 — Experiment Description. NASA TM-78173 (1978)
71. Cogoli, A.: Acta Astronaut. (in print)
72. Cogoli, A., Tschopp, A.: ESA Bulletin 24, 24 (1980)
73. Cogoli, A., in: Spacelab Mission 1 — Experiment Description. Craven, P. D. (ed.), NASA TM-78173, pp. V-17/V-19 (1978)
74. Schachter, P., Tyler, J.: Life Sciences Laboratory Equipment (LSLE) Description. NASA JSC-16254-B (1980)
75. Study on Incubators and Holding Units for Cells and Tissues, Plants and Lower Vertebrates. ESA CR(P) 943 (1977)
76. Biorack — Report on the Technical Feasibility Study. ESA DP/ST(80) 1 (1980)
77. Spacelab Payload Accommodation Handbook. Issue No. 1, Revision No. 4. ESA SLP/2104 (1980)
78. Safety Policy and Requirements for Payloads Using the Space Transportation System. NASA NHB 1700.7A (1980)
79. ESTEC Test Facilities. ESTEC TTO-315/1 (1979)
80. Cogoli, A., et al.: Life Sci. Space Res. 17, 219 (1979)
81. Cogoli, A., et al.: Aviat. Space Environ. Med. 51, 29 (1980)
82. Cogoli, A., et al.: The Physiologist 22, S-29 (1979)
83. Cogoli, A., Tschopp, A.: The Physiologist 23, S-63 (1980)
84. Cogoli, A., in: Advances in Physiological Sciences — Gravitational Physiology. Hideg, J., Gazenko, O. (eds.), Budapest: Akadémiai Kiadó, Vol. 19, p. 87 (1981)
85. Smith, A. H., Rhode, E. A., Spangler, W. L.: Cumulative aspects of repeated HSG Exposure. Report SAM-TR-79-17, USAF School of Aerospace Medicine (1979)
86. Briegleb, W., Schatz, A., Neubert, J.: Umschau Wiss. Techn. 76, 621 (1976)
87. Silver, I. L.: J. Theor. Biol. 61, 353 (1976)
88. A Forecast of Space Technology 1980—2000. NASA SP-387 (1976)

89. Botkin, D. B., et al.: Life Sci. Space Res. *17*, 3 (1979)
90. Krauss, R. W.: Life Sci. Space Res. *17*, 13 (1979)
91. Spurlock, J. M., Modell, M.: Life Sci. Space Res. *17*, 27 (1979)
92. Brown, A. H., et al.: Life Sci. Space Res. *17*, 37 (1979)
93. Grey, J. (ed.): Space Manufacturing Facilities (Space Colonies). American Institute of Aeronautics and Astronautics, Inc. New York 1977

Basic Concepts in Microbial Aerosols

J. E. Zajic,
College of Science University of Texas at El Paso, El Paso, Texas 79968, USA

I. I. Inculet, P. Martin
Faculty of Engineering Science the University of Western Ontario, London, Ontario, Canada N6A 5B9, Canada

This literature survey contains information pertaining to bacterial aerosol viability and electrostatic surface charges of bacteria. The variables involved in the survival of bacterial aerosols include the method of droplet formation and the intrinsic characteristics of the gas in which the bacteria-containing droplets are suspended. The survey contains information on overall microbial aerosol studies, assay techniques, cell ions, the effect of air composition, the cell surface, radiation effects on aerosols, and the action and the possible components of relative humidity involved in the viability of bacterial aerosols. A small portion of the investigation deals with the survival of mycoplasmas, L-forms, and algae in air.

Also there are sections reviewing the characteristics of aerosol protecting agents, aerosols of *Serratia marcescens* and the pigmentation produced by the bacteria, electrostatic forces relevant to bacterial aerosol deposition, and the changes on cell walls.

1 Microbial Aerosols

A biological aerosol is defined as an artificially generated or naturally occurring collection of biological paricles suspended in air [1]. These particles (bacterial, viral or fungal) can exist in the airborne state as single cells or as clumps of microorganisms as small as 1—10 μm in size. The microorganisms may be attached to dust particles [2] or enclosed in water droplets [3]. Under proper conditions of temperature, relative humidity, and other factors, these living airborne microbes can multiply very rapidly [2].

The ability of airborne microbes to multiply after settling onto an amenable surface is generally considered to be an expression of their viability or survival capability [4]. This survival capability has been directly associated with those bacterial aerosols that have either encountered the least change in their microenvironment [5] or are in the stationary phase of growth (Fig. 1) [6, 7].

There are many sources of bacterial aerosols. Bacteria can enter and be dispersed into the atmosphere as a result of solid waste treatment [6, 8] or from bubbles bursting from the microbially laden surface films of rivers, lakes, and oceans [9], or from spray generated by breaking waves [10, 11] and rainwater splashes [12, 13]. Dislodgement from vegetation or soil as result of wind action or thermal convection [3, 14, 15, 16, 17] is another possible source.

People also disperse microbes by talking, coughing, sneezing [18, 6] and by the shedding of skin [3, 6, 7, 9]. Persons capable of shedding microorganisms from the respiratory tract or other areas of the body in abnormally high numbers (a rate of ≥ 100,000 viable particles per minute, Kraidman [19] are termed microbial 'shedders' and present a constant source of air contamination.

It has been estimated that airborne microbes of 1 μm diameter could be carried 14,000 km (9,000 miles) in a steady 16 km h^{-1} (10 m.p.h.) wind before settling 92 m (100 yards). Bacteria have been isolated at altitudes as high as 256 km (16 miles) [20] to 768 km (48 miles) [21]. Evidently, a generator of pathogenic microbes could exert its effect (contamination or disease) thousands of kilometers away with no apparent link between the terminal settling area and the source. However, the infectivity and viability of airborne microorganisms, carried by the wind, is limited by the availability of substrates [22].

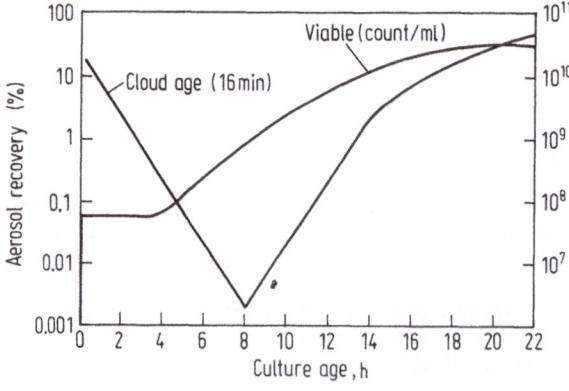

Fig. 1. Aerosol stability of *Serratia marcescens* as a function of culture age (h). (Figure after Goodlow and Leonard [4])

The size of the particle is dependent upon, among other variables the type of microbe, the chemical content of the fluid from which the droplet nuclei originate, as well as the relative humidity (RH) of the atmosphere in which they are suspended [23]. If a particle is less than 10 µm in size, it is defined as a droplet nucleus [24].

Particles over 10 µm in size are almost completely screened out in the upper respiratory passages of man, along with 80% of particles 5 µm in size. Smaller sized particles are inhaled and preferentially deposited in the lung [25], or, if not inhaled, tend to remain suspended in confined and inhabited space until vented or inhaled [26]. Dimmick and Akers [1] state that as droplet size decreases, the surface exposed to air resistance becomes relatively greater when compared to weight or gravitational attraction towards the earth. This ability to remain in static air constitutes a biological hazard in areas of unvented air.

Persons having open wounds, boils, or carbuncles shed microbes of the genera *Streptococcus, Klebsiella* and *Pseudomonas* by air [19]. Other airborne diseases and pathogenic microorganisms are tabulated in Table 1 [27].

The American Association for Contamination Control in 1965 established air quality requirements for hospital rooms, burn areas, dressing rooms and nurseries. For a 60 minute continuously sampled period, the flow of air entering these areas should contain no more than one particle-bearing, colonyforming aerobic bacteria per cubic foot; the air already present within this space should not hold more than 5 particles per cubic foot. However, air conditioners, ventilators, fans, etc. can create aerosols and disperse them throughout a room thereby increasing the probability of contagion inhalation or deposition.

The parameters involved in the assay as well as viability and techniques are reviewed in the following sections. Table 2 summarizes some of the work done in aerosol viability studies. Table 2 lists much of the work in aerobiology identifying and elaborating on the stresses involved in microbial aerosols. The other factors that are probably involved in bacterial aerosols are the physical forces associated with the generation and recovery of the aerosol and the presence of extraneous material such as spent culture fluids or tissue exudates in the aerosol or collecting fluids [28]. The increase or decrease of the death rate of microbes in these aerosols depends upon the quality and/or quantity of the above factors.

Table 1. Pathogenic microorganisms and diseases caused by air transfer

Organisms: Adenoviruses	Diseases: Blastomycosis
Bordetella pertussis	Histoplasmosis
Cryptococcus sp.	Influenza
Diplococcus pneumoniae	Mumps
Mycobacterium tuberculosis	Psittocosis
Mycoplasma pneumoniae	
Neisseria meningitis	
Poliomyelitis virus	
Rubeola virus	
Streptococcus haemolytica	
Variola virus	

Table 2. Factors involved in microbial aerosol viability

Parameters	Author	Ref.
Bacterial strain	Goodlow & Leonard (1961)	4)
	Dimmick & Akers (1969)	1)
	Pelczar & Reid (1972)	27)
Incubation medium	Goodlow & Leonard (1961)	4)
	Dimmick & Akers (1969)	1)
	Cox (1971)	29)
Temperature	Goodlow & Leonard (1961)	4)
	Hayakawa & Poon (1965)	2)
	Akers et al. (1966)	30)
	Hatch & Dimmick (1966)	31)
	Cox (1966a, b)	6, 7)
	Anderson (1966)	28)
	Wright et al. (1969)	32)
	Dimmick & Akers (1969)	1)
	Ehrlich et al. (1970)	33, 34)
	Ehrlich & Miller (1973)	35)
	Cox et al. (1974)	36)
Relative humidity	Dimmick (1960)	5)
	Hayakawa & Poon (1965)	2)
	Hess (1965)	37)
	Akers et al. (1966)	30)
	Hatch & Dimmick (1966)	31)
	Won & Ross (1966)	38)
	Cox (1966a, b)	6, 7)
	Anderson (1966)	28)
	Zentner (1966)	39)
	Cox (1968a)	40)
	Wright et al. (1968)	41)
	Dimmick & Akers (1969)	1)
	Stewart & Wright (1970)	42)
	Ehrlich et al. (1970b)	34)
	Hatch et al. (1970)	43)
	Southey & Harper (1971)	17)
	Riley & Kaufman (1972)	44)
	Cox & Goldberg (1972)	45)
	Lighthart (1973)	46)
	Turner & Salmonsen (1973)	47)
	Ehresmann & Hatch (1975)	48)
	Cox (1976)	49)
Cellular physiology	Hatch & Dimmick (1966)	31)
	Anderson & Cox (1967)	50)
Oxygen concentration	Hess (1965)	37)
	Cox (1967; 1968a)	51, 40)
	Benbough (1969)	52)
	Cox et al. (1974)	36)
	Cox et al. (1973)	54)
	Cox et al. (1971)	53)
Light and irradiation	Kundsin (1966; 1968)	23, 26)
	Serat et al. (1969)	57)
	Lighthart et al. (1971)	58)
	Cox et al. (1973)	59)
	Lighthart (1973)	46)

Table 2 (continued)

Parameters	Author	Ref.
Ion transport disruption	Anderson et al. (1968)	60)
	Hambleton (1971)	61)
	Benbough et al. (1972)	62)
Sampling medium	Anderson (1966)	28)
	Cox (1966a, b)	6, 7)
	Hatch & Dimmick (1966)	31)
	Won & Ross (1966)	38)
	Kethley et al. (1957)	63)

2 Assay Techniques

Many nebulizers are obtainable that generate aerosols from solutions depending on the particle size, uniformity of droplets, and dispersion conditions required. Information on the types and characteristics of atomizers has been reported by Mercer [64], Dimmick and Akers [1] and Larson [65].

Dimmick and Akers [1] suggest a direct way to measure the output of cells from a nebulizer by straining the cells, dispersing them with distilled water, collecting them in a sampler (for further information see Table 3), and measuring the concentration by photometric methods. The volume (ml) of the dispersed droplet times the number of cells per ml yields the expected probability of a cell being in a droplet.

The number distribution of aerosolized particle sizes is approximately log normal [66] except for the smallest particle size. If higher numbers of cells are being contained in the smaller particles [4], the concentration of cells in the dispersing fluid is increased and the solid content as well as the particle density increases in the atomizer fluid. This effect is negligible for spray times less than 5 min [1].

The decay of a microbial aerosol can be estimated by quantifying the number of airborne organisms per unit of air volume at various cloud ages. The slope of the resulting curve (density vs cloud age) is the rate of decay of the aerosol cloud (% per min). Included in this decay rate are physical decay caused by fallout or deposition on the chamber walls and biological decay caused by biological inactivation of the microbes [33].

Table 3. References for the collection and sampling equipment used in aerosol studies

	Ref.		Ref.
Luckiesh, Taylor and Holladay (1946)	67)	Andersen and Andersen (1962)	74)
Cown, Kethley and Fincher (1957)	68)	Lidwell and Noble (1965)	75)
Houwink and Rolvink (1957)	69)	Malligo and Idoine (1964)	76)
Andersen (1958)	70)	May (1964)	77)
Decker et al. (1958)	71)	Mercer, Goddard and Flores (1965)	64)
Wolochow (1958)	72)	Dimmick and Akers (1969)	1)
Batchelor (1960)	73)	Knuth (1969)	78)

Recovery of a tracer disseminated as a mixture with the test microbes allows an estimation of the physical decay of the aerosol. Therefore, the difference between the total decay rate and the physical decay rate provides an estimate of the biological decay of the organisms. Hatch and Dimmick [31] state that, while physical decay always follows first order kinetics, the biological decay rate, depending on the conditions prevailing, only approximates first order kinetics.

Cox et al. [79] provide equations to estimate the components (electrostatic precipitation, inertial impaction, gravitational settling, and Brownian diffusion) of the aerosol physical decay. A phenomenon of the decay rate that must be taken into account during aerosolization is the extremely high decay rate in the first second of the cloud age which is followed by a slower rate of decay during the next nine seconds. This means that the decay rate is not necessarily constant with respect to time.

Assuming first order kinetics, the change in the number of aerosol particles, dN, is equal to the change in time of the suspended particles, dt, times a constant rate k, and the microbe concentration N,

$$\text{i.e. } \frac{dN}{dt} = -kN \tag{1}$$

Rearranging and integrating Eq. (1) from $t = 0$ to some time t yields

$$\int_{N_0}^{N_t} -\frac{dN}{N} = \int_0^t k \, dt \tag{2}$$

$$-\int_{N_0}^{N_t} \frac{dN}{N} = kt \tag{3}$$

Integrating and solving for k gives

$$k = \frac{1}{t} \ln \left(\frac{N_0}{N_t} \right) \tag{4}$$

where k = decay rate constant, s^{-1}
       t = time of sampling, s
       N_o = original number (or concentration) of aerosol particles (microbes), particles-cm^{-1}
       N_t = number (or concentration) of aerosol particles (microbes) at sampling time t, particles-cm^{-1}

Hayakawa and Poon [2] used Eq. (4) to derive the decay rate constant for their data. This is not the only approach to evaluating microbial activity in aerosols. For biological decay, Hatch and Dimmick [31] followed the change in relative humidity (RH) from the initial cloud conditions to the final cloud conditions and correlated the biological loss of the aerosol to a dynamic-humidity-death (DHD) ratio (Fig. 2).

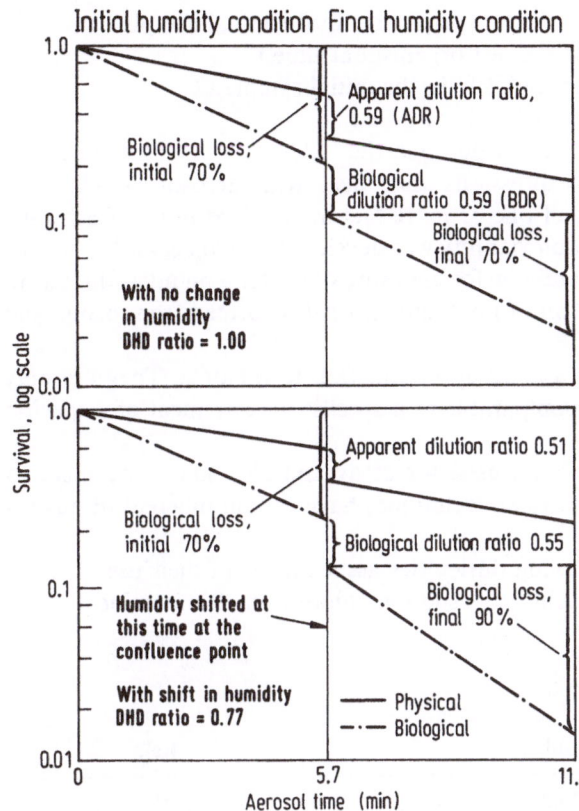

Fig. 2. The change in survival rate with a change in the relative humidity of suspended bacterial aerosol particles. The physical loss is due to natural forces (e.g. coagulation, deposition, etc.); the biological loss refers to the loss of viability. (Figure after Hatch and Dimmick [31])

Other quantitative methods relate the colony forming units (CFU) of the experimental microorganisms to the CFU of a microbial tracer (spores) to yield the percent viability (or percent survival; Cox and Goldberg [45]):

$$\% \text{ viability} = \frac{\text{CFU (test organism)}}{\text{CFU (tracer organism)}} \times 100^* \tag{5}$$

Lighthart [46] employed a graph of $-\log (\% \text{ survival})$ versus time interval and states that the death rate is the negative slope of the line, fitted by the method of least squares through the experimental points. The equation obtained from this is

$$
\begin{aligned}
\% \text{ survival} &= \frac{\text{total decay}}{\text{physical decay}} \times 100 \\
&= \frac{(N_{UI}/N_{UT})}{(N_{BT}/N_{BI})} \times 100
\end{aligned}
\tag{6}
$$

---

* The ratio of CFU's was normalized to the initial CFU ratio in the aerosol medium to obtain a comparative reference point to 100% viability.

where N_{UI} = number of CFU of the test organisms in the initial sample

N_{UT} = number of CFU of the test organism at time t

N_{BI} = number of tracer spore CFU in the initial sample

N_{BT} = number of tracer spore CFU at time t.

The previous equations were used to describe the "life-times" of bacteria in the aerosol state. Other algebraic expressions that deal with aerosol viability are mainly concerned with the effect of one or more factors involved in the death rate. An equation for the oxygen effect on viability was developed by Cox et al. [36], one for a denaturation effect by Cox [49], and one for exposure of bacteria on microthreads to an open air factor effect (the action of pollutants probably formed from ozone and olefin interaction) by Cox et al. [59].

All of these formulations give a good approximation to the data. They are only approximations and may not be comparable to a specific measurement of a specific microbial aerosol.

Table 4 includes the tracer material used for assay and also some references to authors using each method. All of these techniques have certain inherent disadvantages.

Chemical tracers have the following difficulties associated with their use. Tracers of relatively small molecular weight (e.g. sodium fluorescein, Rhodamine B or a

Table 4. Assay tracers and experimental use

Tracer or Isotope	Author	Ref.
Dyes	Dunklin & Puck (1948)	81)
	Henderson (1952)	82)
	Wolfe (1961)	83)
	Dorsey et al. (1970)	55)
^{32}P	Harper & Morton (1952)	84)
	Harper et al. (1958)	85)
	Harper (1963)	86)
	Harstad (1965)	87)
	Green & Green (1968)	88)
	Benbough (1969)	52)
^{35}S	Miller et al. (1961)	89)
	Green & Green (1968)	88)
^{14}C	Anderson (1966)	28)
	Benbough et al. (1972)	62)
^{125}I	Strange et al. (1971)	90)
	Strange & Martin (1972)	91)
	Strange et al. (1972)	92)
Bacillus subtilis var. *niger* spores	Anderson (1966)	28)
	Cox (1966a, b; 1968a, b; 1971; 1976)	6, 7, 40, 93, 29, 49)
	Benbough (1969)	52)
	Ehrlich et al. (1970)	34)
	Hambleton (1970)	94)
	Strange et al. (1972)	92)
	Cox & Goldberg (1972)	45)
	Ehrlich & Miller (1973)	35)
	Lighthart (1973)	46)
	Cox et al. (1974)	36)

radioisotope) are evenly distributed within the aerosol medium and particles. However, unless the tracer is specific for the microbe, the recorded count will vary because some of the aerosol particles do not contain microbes. In other words, the difference in the physical properties of the particles affects the collection ratio of tracer to microbial particles. Radioactive tracers have the same difficulties but, in addition, safety must be considered.

Viability problems are encountered in the use of spores as tracers. There is evidence that a proportion of the spore tracer may die in the aerosol [85, 80, 28, 7, 34]. Strange et al. [92] have been shown that about 10% of the initially viable spores (*Bacillus subtilis* var. *niger*) lost viability or became dormant under certain conditions in an aerosol. This loss of tracer CFU ability imparts a higher survival or viability percentage and a lower death rate to the test microbes than is warranted.

Overall, assays of different bacterial aerosols generally have shown that Gram negative microbes are more sensitive to aerosolization than are Gram positive organisms [95, 80, 96, 88]. Goodlow and Leonard [4] state that thermophilic mutants of several pathogens could possibly show increased resistance to aerosolization. Finally, despite the great physiological and structural differences between the true bacteria, mycoplasmas, and L-forms, these species show similar responses to the stress of being in the airborne state [26, 41, 32, 42].

3 Concept of Cell Ion Units

The permeability of some Gram negative bacteria is affected by aerosolization. Lack of control of ion transport [60] and increased permeability [98, 99] have been associated with microorganisms recovered from aerosols. In addition, there is a change in the outer layers of the cell envelope [94]. Consequently, organisms which have been recovered from aerosols should not be regarded as unchanged rehydrated forms of the original bacteria. Permeability changes resulting from aerosolization may not be lethal in themselves but could be directly or indirectly related to the subsequent survival of airborne bacteria [60, 94, 61].

Damadian [100] hypothesized that *E. coli* can be viewed as simple, multiphase ion-exchange systems and that the movement of ions such as potassium (K^+) and sodium (Na^+) into the cells during growth and respiration are only counterions for newly formed polymers or acid products of catabolism. However, Anderson and Dark [101] think that damage to a K^+ retention system is a consequence of the aerosolization of *E. coli*. In fact, it is the general opinion [28, 101, 52, 94] that K^+ retention is lost by most aerosolized bacteria.

Aerosolization of *E. coli* decreases their ability to synthesize certain macromolecules and concentrate ions [60, 52, 62]. Anderson et al. [60] state that the loss of control of K^+ retention results in a general disorganization of the bacterial ion and substrate transport mechanism. However, the increased K^+ permeability can be counteracted by the addition of K^+ in the growth medium.

Anderson [28] established that populations of *E. coli* B showed a severe decrease in the ability to synthesize protein immediately after aerosol sampling. Since there is evidence that ribosomal function depends on K^+ in a mutant of *E. coli* and in *Aerobacter aerogenes* [103], the decreased rate of protein production could be due to the loss of K^+ control. Inasmuch as K^+ is a cofactor for many enzymes in carbo-hydrate and aromatic compound metabolism [101], these reactions could also be affected by a disturbance in the K^+ retention system.

Benbough et al. [62] found that aerosolization caused a detachment of bacterial membrane components. The transport mechanism was restored by incubation of the microbes with the lost components. Increased K^+ permeability and the loss of incubation protein synthesis are also thought to be reversible [28, 104, 101].

On the other hand, experimental work with ^{32}P has illustrated that much of the loss of the phosphate ion from the cell has been due to procedural methods previous to aerosolization [101]. It appears that the process or procedure for atomization or a stress involved in being suspended in air can cause a partially reversible and partially lethal loss of at least one ionic component in aerosolized bacteria.

4 The Effect of Air Composition on Microbial Aerosols

4.1 Oxygen

The effect of oxygen on airborne organisms are complex and could either be toxic or protective. A protective action of oxygen was hypothesized to occur through a suppression of the activity of a toxic component in the spent culture suspending medium [29]. Conversely, if *Serratia marcescens* were held in an oxygen free atmosphere, there was little or no viability loss [37] i.e. the toxic effect of oxygen was absent. There appear to be similarities in the toxicity of oxygen towards aerosolized bacteria and its effect on enzyme activity [105], irradiated organisms [106, 107, 108], and freeze-dried organisms [109, 110].

Cox et al. [36] state that aerosolization and freeze-drying can both result in the same oxygen induced death mechanism. They derived a kinetics equation for the viability of freeze-dried bacteria to fairly accurately describe the survival of aerosolized bacteria in varying oxygen concentrations. Cox and Baldwin [111] suggest that the oxygen toxicity towards aerosolized bacteria may be a free-radical induced phenomenon. Freeze-dried *Escherichia coli* have shown electron spin resonance signals [112, 113, 109] which are indicative of free-radical formation. These signals were greatly enhanced on exposure of the freeze-dried microbes to oxygen.

Since oxygen induced free-radicals decay rapidly in the presence of moderate amounts of water vapour [114], this toxicity can occur at low relative humidity (e.g. low survival of *Pasteurella tularensis* LVS at low RH values, Cox [29]. If oxygen toxicity occurs at high and intermediate RH values (see Table 2) [6, 40, 52] another process must be involved in the death mechanism.

Oxidation by peroxy radicals (from exposure to oxygen) at low RH is a consequence of bacterial dehydration. In other words, the lethal action of oxygen appears to be associated with the loss of water from bacteria. Oxygen was shown to be progressively more toxic as the humidity is reduced below 65% [52]. Bateman et al. [115] pointed out that perhaps it is the strongly bound water that is lost when bacteria are exposed to RH values of 70% and less. The structural conformation of bound water could alter oxygen accessibility to the bacteria and its reactivity with bacterial macro-molecules [80].

Oxygen has been found, at atmospheric pressure, to produce enzymatic inhibi-tions [116]. Specifically, hyperbaric oxygen oxidizes the dithiol moiety of a-lipoic acid, a cofactor of pyruvate oxidase which, in effect, decreases enzyme activity in brain cells [117]. Barron [118] and Hangaard et al. [105] have also observed the toxic action of oxygen on pyruvate oxidation in gonoccoci and heart homogenate, respectively. Although it has not been demonstrated, this action of oxygen on enzyme systems could occur in aerosolized bacteria.

The loss of the outer membrane components of aerosolized Gram negative bacteria by the process of aerosolization, as described previously, could expose microbial proteins to the toxic effect of oxygen. The sensitivity of the microorganisms in aerosols would therefore increase. Cox [40, 29] suggests that the toxic action of oxygen is on the inhibition of cell division, or cell synthesis, or both. The site of the oxygen reaction could be in the periplasmic gap [36]. Goodlow and Leonard [4] found that minimal bacterial survival in air is obtained with rapidly metabolizing S. marcescens. Gershmann et al. [119] also noted that protection against oxygen poisoning is the greatest during decreased metabolism of irradiated microbes.

Oxygen toxicity can be altered by changing the growth conditions or by including certain agents (e.g. Mn^{+2}, Co^{+2}, glycerol and thiourea) in the spray fluid. These agents also increase the viability of freeze-dried bacteria [52].

Physical forces, water content, metabolism, medium constituents and oxygen concentration are the variables that should be included in any study of the effect of oxygen on microbial aerosol survival.

4.2 Other Gases

Bacterial aerosol survival in air can be similar to (at high RH) or much less than (at low RH) survival in pure nitrogen [7]. Freeze-dried organisms also show an increased survival in atmospheres of nitrogen, hydrogen, and carbon dioxide when compared to air or oxygen [120, 121]. However, Webb [80] found that the viability of aerosolized bacteria was not different in air than in other gases.

Carbon monoxide can have a protective or lethal effect on aerosolized bacteria in concentrations present in the urban environment. The death rate of Serratia marcescens and Sarcina lutea was enhanced by carbon monoxide (85 µl per 1 air) at low RH. Cell protection was shown at high RH values. The microbial species, aerosol age, and RH are some of the factors that have been studied by Lighthart [46] in conjunction with carbon monoxide as a contaminant.

In some organisms, cellular metabolism is affected by a carbon monoxide-cytochrome reaction resulting in photosensitive carbonyl compounds [122, 123]. It has also been shown that bacterial chromosome replication depends upon a carbon monoxide sensitive reaction which in itself affects bacterial survival [124]. Lighthart [46] explained his results from the viewpoint of the inhibition of an energy-requiring death mechanism at high RH and an inhibition of energy necessary for cell maintenance at low RH values.

Lighthart et al. [58] used sulphur dioxide as a variable and concluded that the death rate of S. marcescens increases directly with gas concentration and inversely with RH. The open air factor (ozone-olefin interaction), in the parts per hundred million range, is very detrimental to microbial aerosol survival [59]. Cox et al. [59] advocate the use of bacterial sensitivity as an indicator of the concentration of contaminants.

Other pure gaseous atmospheres of nitrogen, argon, and helium have been used in aerosol studies to exclude the detrimental action of oxygen on the airborne microbes. Cox [40] found, at high RH, that these gases were biologically active and possibly formed clathrates with cellular water. These interactions will be discussed briefly in the section concerning RH.

5 Cell Surface

The cell surface is affected by almost all aspects of the aerosolization process (e.g. surface charge, binding and adhesion, amount of bound water, reaction with inert air pariculates, etc.). Most bacterial and plant cells are surrounded by a wall, consisting of a thick porous, polymeric meshwork that may occupy as much as 50%

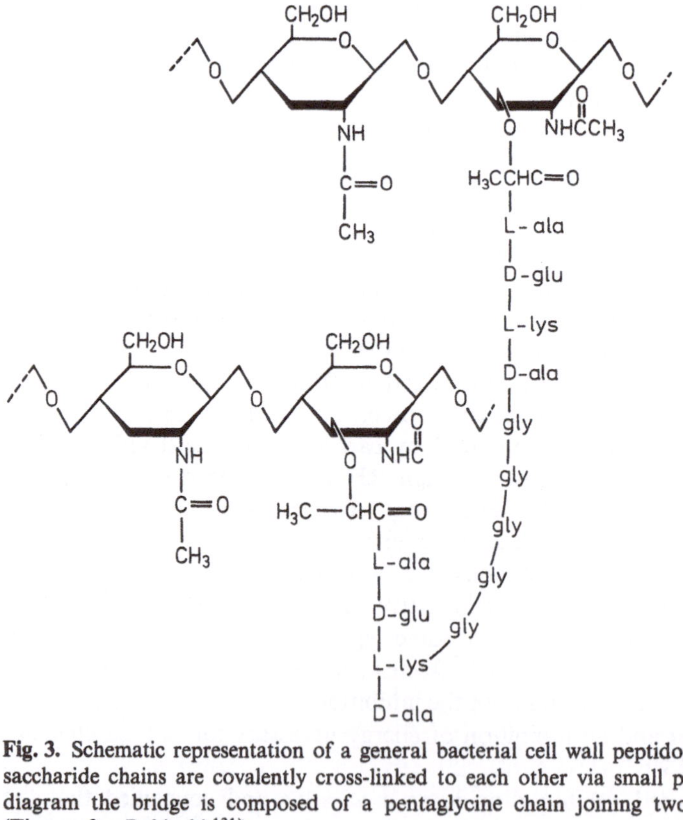

Fig. 3. Schematic representation of a general bacterial cell wall peptidoglycan complex. The polysaccharide chains are covalently cross-linked to each other via small polypeptide bridges. In this diagram the bridge is composed of a pentaglycine chain joining two tetrapeptide side chains. (Figure after Bohinski [131])

of the total volume [125]. The surface structures of a variety of bacterial cells are extensible, flexible, and elastic as shown by the study of Knaysi [126].

Electron microscopic studies [127] have shown that Gram negative bacterial cells consist of an inner rigid peptidoglycan layer and a membrane-like outer 'soft' layer. The peptidoglycan moiety comprising a part of the cell wall structure can be visulaized as a 3-dimensional rope ladder with relatively rigid polysaccharide rings and relatively flexible polypeptide ropes (Fig. 3, 4 and 5). Ou and Marquis [125] describe this network as a predominantly elastic restraining structure rather than a rigid shell.

Figure 5 shows the main constituents of the *Escherichia coli* cell envelope and the sites of action of three enzymes. The flexible lipoprotein-lipopolysaccharide component [123, 129] that overlays the layer of mucopolysaccharide (peptidoglycan) protects this layer from the action of lysozyme, possibly by impeding the penetration of the enzyme into the wall [94]. In support of this theory, Salton [128] and Noller and Hartsell [130] demonstrated that certain pretreatments, which appear to act by disruption of the outer wall complex of Gram negative bacteria, sensitize these bacteria to lysozyme.

The effect of aerosolization on the cell envelope may be non-specific [60]. Hambleton [94] states that, following aerosolization of bacteria in air at 75 % relative humidity (RH), some Gram negative bacteria show structural changes in their envelopes, manifested in the susceptibility of the organisms to some enzymes, in particular, lysozyme.

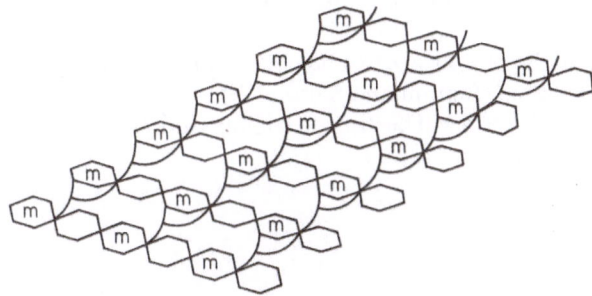

Fig. 4. Schematic diagram of the peptide-carbohydrate repeating units (Fig. 3) repeated to yield a rigid, gridlike structure. The 'm' refers to the carbohydrate that is involved in the cross-linking (Figure after Bohinski [139])

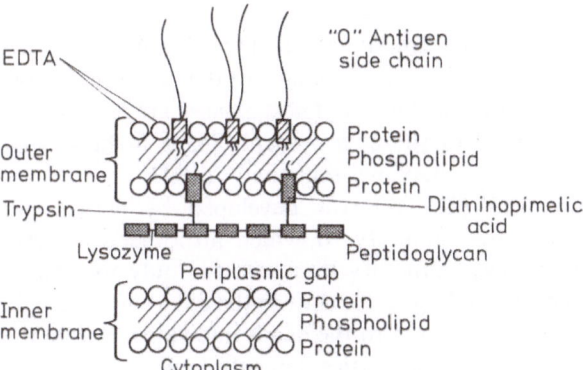

Fig. 5. Simplified model of the cell wall envelope of *Escherichia coli*. The underlined (with broken lines) compounds are enzymes and the arrows indicate the site of the enzymatic activity. (Figure after Schnaitman [132])

Hambleton [61] found that certain di- and trivalent cations added to suspensions of bacteria recovered from aerosols mitigate this lysozyme sensitivity. Aspell and Eagon [133] and Schnaitman [132] seem to agree that the role of divalent cations is in the stabilization and organization of the lipopolysaccharide-lipoprotein layers of the Gram negative cell envelope. Therefore, since sensitivity to lysozyme does occur and divalent cations mediate a resistance to the enzyme, aerosolization must stress or structurally affect the permeability of the outer membrane to lysozyme (Fig. 5).

Hambleton [61] also states that the repair of wall damage of aerosolized bacteria is similar to the restoration of permeability control occurring in organisms recovering from EDTA (ethylene diaminetetraacetic acid) treatment [134, 132]. This hypothesis can be easily tested by analytic comparison of the results of the two treatments.

Lieve [135] has demonstrated that the changes in permeability inducted in E. coli by EDTA can be reversed by an energy utilizing process in the microbe. Maximum wall repair of aerosolized bacteria has been shown to take place only when the microorganisms were incubated in a complete medium in the presence of multivalent cations [61]. On the other hand, aerosolized bacteria, in the presence of 2,4-dinitrophenol, which uncouples oxidative phosphorylation [136], are unable to restore lysozyme resistance (i.e. to repair the outer membrane; Hambleton [61].

It is reported that mycoplasmas or L-forms form viable aerosols even though they lack a cell wall [23, 26, 42].

Preferential removal of surface cell wall components is possible. In E. coli, ethanol removes the lipoprotein of the cell wall to expose the mucopeptide layer [137, 138].

Enzymatic removal of the carboxyl terminal groups of surface amino acids was obtained by incubating E. coli under certain conditions and harvesting a specific decarboxylase [139].

Blockage of the amino groups on the cell surface can be obtained by treatment with hyaluronidase and fluorodinitrobenzene (Ingram and Salton [140]).

Osmotic variations of aerosolized bacteria will be discussed in the section concerning relative humidity and water effects in aerosolization.

6 Relative Humidity

6.1 General

Water is an ubiquitous compound that is essential to all life. The liquid phases surrounding and dominating the inside of a bacterial cell govern the rate of transport of nutrients, ions and gases, into and out of the cell. These phases are sensitive to variations in the relative humidity (RH) of the environment. Table 5 shows that various aerosolized bacteria have ranges of maximal sensitivity to RH.

A dual aerosol transport apparatus (DATA) was developed by Hatch and Dimmick [31] to adiabatically change the humidity to which airborne cells were initially equilibrated to a higher or lower humidity level and to study the effects of humidity shifts. An airborne microbe undergoing a shift in the environmental RH will have an increased death rate when compared to the death rate values before the shift. The modification of cellular biological properties depends on the direction of

Table 5. References for aerosol experimental results supporting different RH ranges of greatest inactivation

Low RH ($\simeq$ <20%)	Mid RH (20–60%)	High RH ($\simeq$ >60%)
Wells & Zapposodi (1948) [141]	Dunklin & Puck (1948) [81]	Williamson & Gotaas (1942) [146]
Dimmick (1960)[a] [5]	Schechmeister & Goldberg	Loosli et al. (1943) [147]
Goodlow & Leonhard (1961) [4]	(1950) [142]	DeOme (1944) [148]
Hayakawa & Poon (1965) [2]	Ferry & Maple (1954) [143]	Dimmick (1960)[a] [5]
Lighthart (1973) [46]	Beebe (1959) [144]	Goodlow & Leonard (1961) [4]
Cox, Gagen & Baxter	Webb (1959) [80]	Cox (1968a) [40]
(1974) [36]	Anderson, Dark & Peto	Ehrlich, Miller & Walker
	(1968) [60]	(1970b) [34]
	Wright, Bailey & Hatch	Cox (1976) [49]
	(1968a) [41]	
	Hatch & Wolochow (1969) [145]	
	Wright, Bailey & Goldberg	
	(1969) [32]	
	Stewart & Wright (1970) [42]	
	Cox (1976)[b] [49]	

[a] different with different atomizing fluid
[b] two ranges of RH hastens decay rate
[c] different for different organisms
[d] survival descreases with increasing humidity

the shift and on the number of times the cells are subjected to a changing environment [43].

Freeze-dried organisms undergo an "RH" shift when allowed to equilibrate to a higher environmental RH. Lyophilized cells of *Serratia marcescens* die rapidly at a water content similar to the equilibration to water vapour at an RH of approximately 90% [149, 150, 115]. Also, the decay rate of freeze-dried aerosolized *S. marcescens* increased at a high RH relative to the decay at low RH. However, a reverse relationship was found when these organisms were aerosolized from a liquid menstrua, e.g. heart infusion broth [5].

Similar unstable (low survival) zones have been found for dried *Escherichia coli* in vacuo at high RH [152] and for aerosolized *E. coli* at RH values of 100, 87, 85 and 50% [28]. Cox [49] states that freeze-drying of bacteria to different water contents is analogous to aerosolization into atmospheres of different RH values (they have similar viability equations).

Cox [6, 7, 51] and Cox and Baldwin [111] found that the survival of *E. coli*, after being aerosolized into a high RH environment, was sensitive to RH in a manner dependent upon the content of the spray fluid (Figs. 6 to 7). It has also been reported that the effect of an RH change before collection depended upon the strain of *E. coli* aerosolized and the nature of the spray and collecting fluids [7, 51].

However, aerosol clouds of *Pasteurella tularensis*, at low to middle range humidities (20—50%), showed a continuous biological decay rate independent of the type of collecting fluid used [38]. Generally, the behavior of aerosolized bacteria to RH differs with the dispersing fluid (Figs. 8, 9, 10) [6, 7, 29, 52, 45], sampling (assay) medium (Figs. 11 and 12) [6, 40, 93], atmospheric components (Figs. 13 and 14) [6, 40, 46], temperature [153], and aerosol age (Fig. 15) [6, 101].

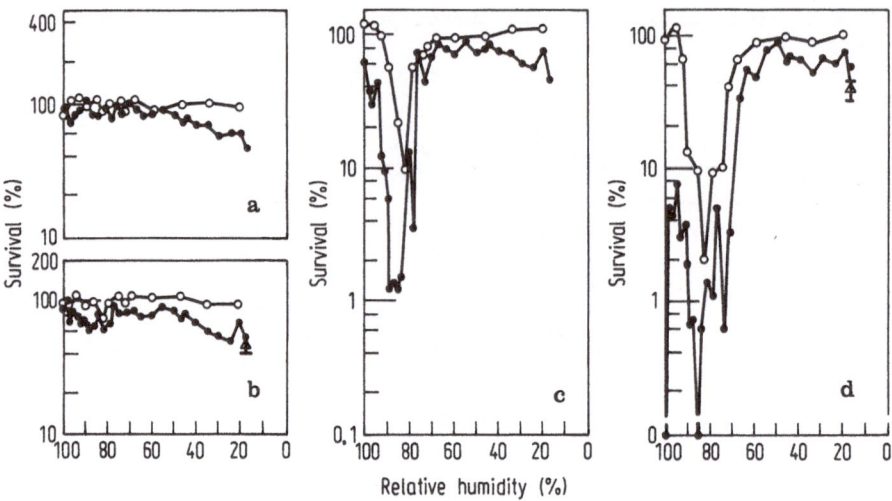

Fig. 6. Survival of *Escherichia coli* (Jepp) in nitrogen at aerosol ages of **a** 0.3 s, **b** 3 s, **c** 2 min, and **d** 15 min. The solid dots are the results for bacteria sprayed from distilled water. The circles are the results for bacteria sprayed from 0.13M-raffinose solution. The bacteria were collected in phosphate buffer. (Figure after Cox [7])

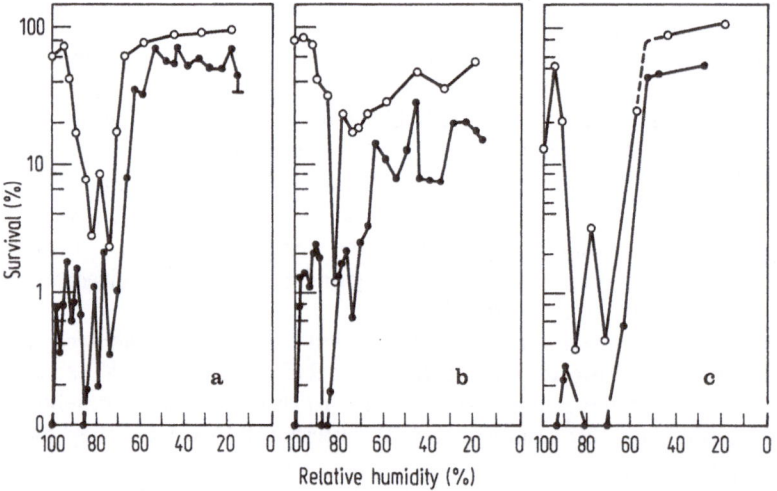

Fig. 7. Survival of *E. coli* (Jepp) in nitrogen at aerosol ages of **a** 30 min, **b** 31.5 min, and **c** 3 h. The symbols are the same as in figure 6. Bacteria are collected in phosphate buffer for **a** and **b**. 1.0 M sucrose in phosphate buffer was used as the collection medium for **b**. (Figure after-Cox [7])

6.2 Temperature

The temperature of the environment plays a role in the stability of airborne organisms. An airborne particle in equilibrium with RH, at a temperature T_1 is not in the same condition as one in equilibrium with RH at T_2 [80] because of the difference in the activation entropy. The activation energy associated with the death mechanism

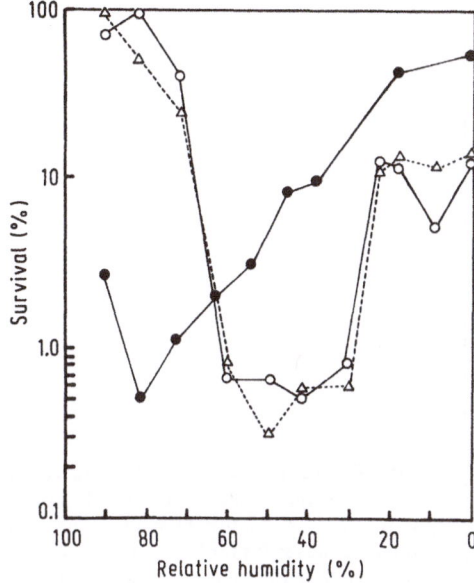

Fig. 8. Survival in air of *Pasteurella tularensis* LVS.

Symbols: o   disseminated from the culture medium;

● disseminated by air in a freeze-dried state (powder);

Δ disseminated from distilled water.

(Figure after Cox and Goldberg [45])

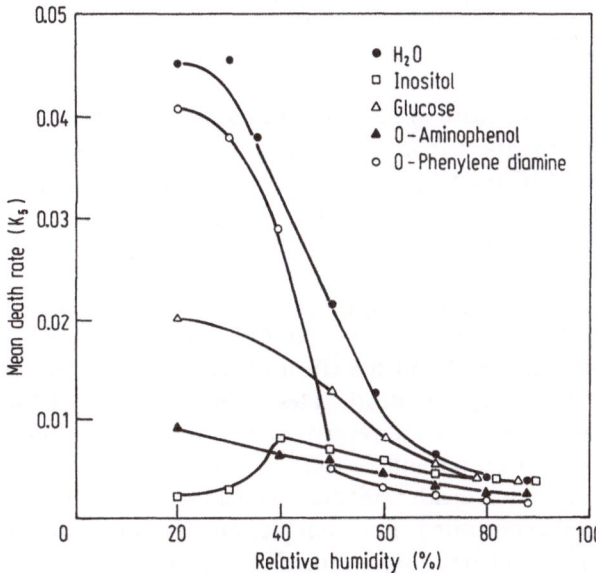

Fig. 9. The effect of relative humidity on the death rate of *Serratia marcescens* aerosolized from various solutions (see symbols on figure). (Figure after Webb [22])

Legend on figure:
● H_2O
□ Inositol
Δ Glucose
▲ O–Aminophenol
o O–Phenylene diamine

and the activation entropy for the decay of the bacterial aerosol can be calculated by plotting the log k (death rate) against the reciprocal of the absolute temperature to yield an Arrhenius plot.

Webb [80] showed from this plot that the decay rate (both in the 0—1 s and the 1—10 s sampling intervals) appeared to follow first order kinetics. For each 10 °C increment of temperature the death rate of *Micrococcus candida* was found to increase 2 to 3 fold [95, 154]. This is similar to the Q_{10} for most biological processes. Other results

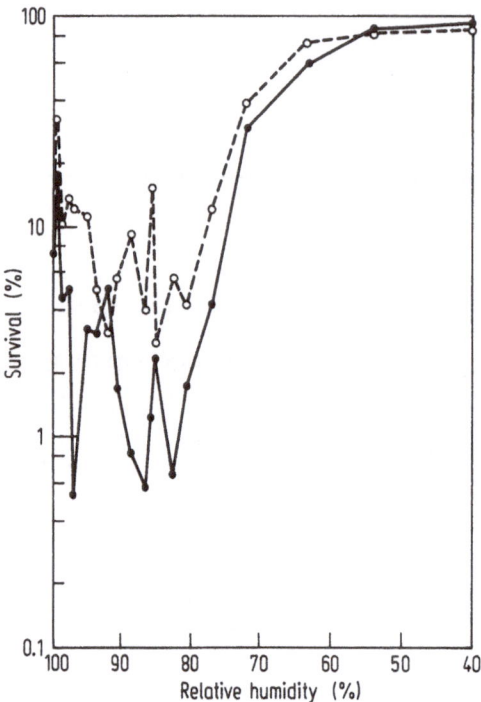

Fig. 10. The aerosol survival of *E. coli* B sprayed from distilled water into helium as a function of relative humidity. The solid dots are the results for bacteria collected into phosphate buffer. The x's are the results for bacteria collected into phosphate buffer + M-surcose. (Figure after Cox [40])

show that, at a single RH value, an increase in temperature resulted in a decreased survival time. For example, the death rate of *P. tularensis* increased with an increasing environmental temperature from 24 °C to 35 °C [35]. *Flavobacterium* sp. [34] *S. marcescens*, and *E. coli* [34] show an increase in the aerosol death rate for a progressive increase in temperature from −18 °C to 49 °C. The latter death rates appear to be linear with temperature.

At low temperatures (−40 °C to 3 °C), the death rate for various airborne microbes was either not noticeably influenced or decreased. At −40 °C, negligible death rates for aerosolized *S. marcescens* were reported and the effect of RH ranging from 20 to 80% was absent [155]. Below 0 °C, very low death rates have been reported for *S. marcescens* [80] and *Flavobacterium* sp. [33]. However, reduced survival has been found for *P. tularensis* and for *S. marcescens* and *E. coli* [35, 34] at 40 °C.

At high temperatures (49 °C), optimum death rates and minimal recovery viability for various organisms have been demonstrated. For example, *Flavobacterium* sp. at high temperatures (>32 °C), when compared to data for *S. marcescens* and *E. coli*, had approximately 1/3 the death rate of the other two vegetative organisms although this was a maximum death rate for this species [34]. At this temperature, thermal inactivation could occur which could lead to alterations in proteinaceous structures and enzymatic activities.

In summary, the factors sensitive to the variation of temperature appear to include the physical characteristics of the slurry, the configuration of the particles upon dissemination [35], temperature sensitive biochemical reactions, varying evaporation rates, the water response to temperature and RH, and structural modifications of bacterial components caused by temperature changes.

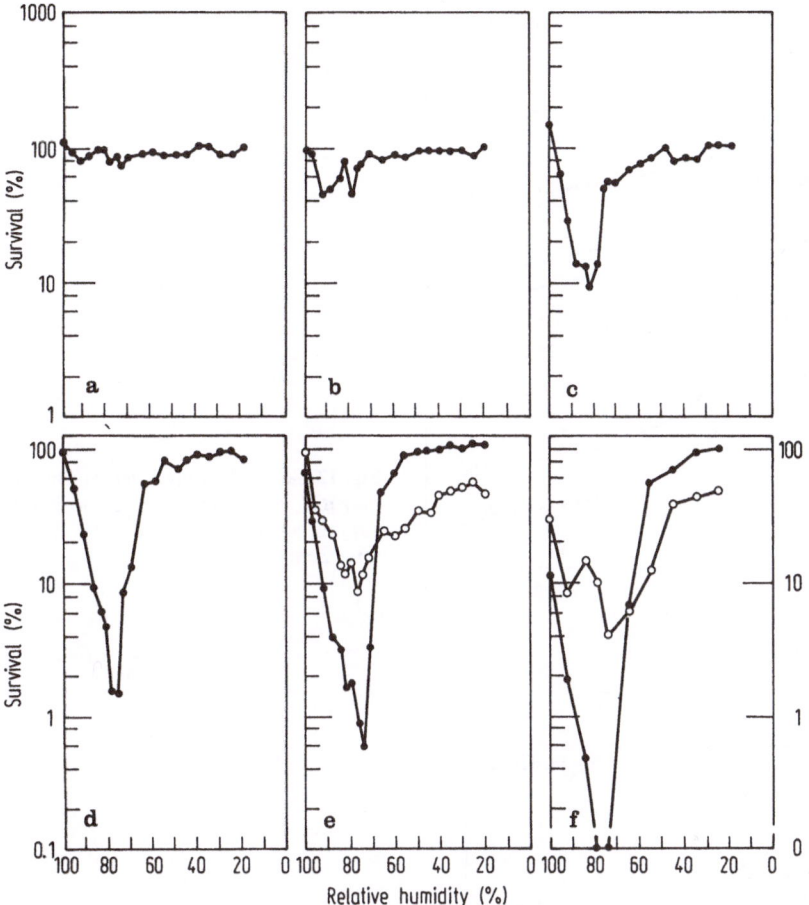

Fig. 11a—f. Survival of *E. coli* (Jepp) in nitrogen sprayed from 0.3 M-raffinose at aerosol ages of **a** 0.3 s, **b** 3 s, **c** 2 min, **d** 15 min, **e** 30 min, and **f** 3 h. The solid dots are results for bacteria collected in phosphate buffer. The circles are the results for bacteria collected in M-sucrose. (Figure after Cox [7])

6.3 Evaporation

Evaporation is dependent on RH, temperature, and the composition of water surrounding airborne particles. Orr and Gordon [156] found a slight decrease in size of individual airborne bacteria with decreasing RH. For a range of bacteria, the viable decay of airborne cells occurred more rapidly in smaller particles than in larger ones [17].

A water droplet of 13 μm diameter, at room temperature, was calculated to evaporate to 0.9 μm in diameter in 0.4 s [2]. Owing to the enhanced vapour pressure over an aerosol droplet surface, the water of an airborne droplet will evaporate even in saturated air through thermal distillation [6]. Water movement out of the particle and intracellular environment may contribute to microbial aerosol death [2, 39]. The evaporation of water from an aerosol droplet results in a lowering of

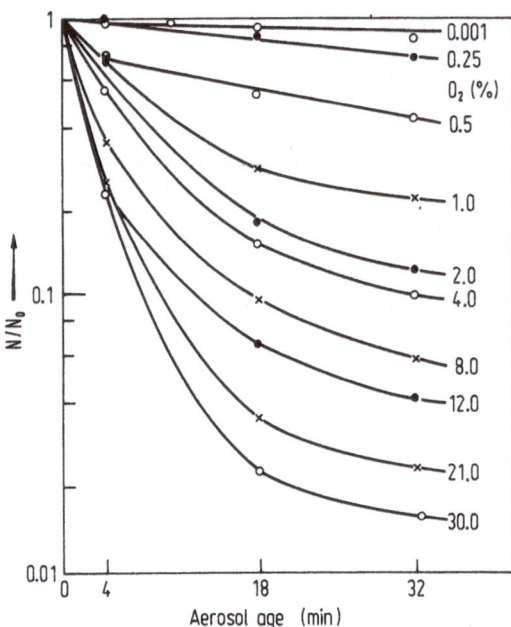

Fig. 12. Aerosol survival versus time for *Serratia marcescens* disseminated into varying concentrations of oxygen. (Figure after Zentner [39])

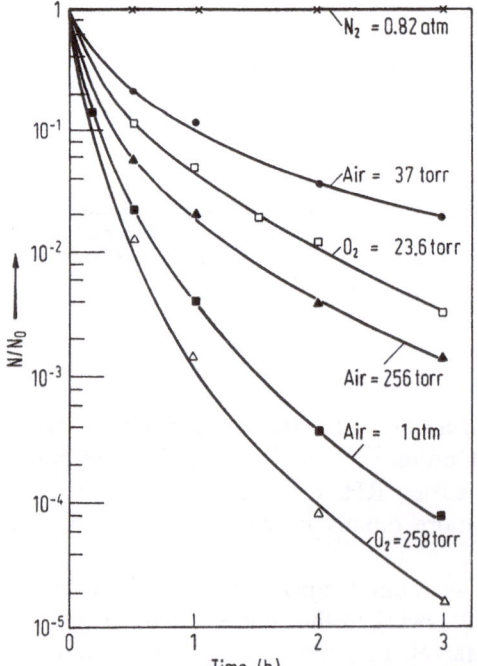

Fig. 13. Survival of *Serratia marcescens* aerosolized into various pressure of air, oxygen, and nitrogen. (Figure after Zentner [39])

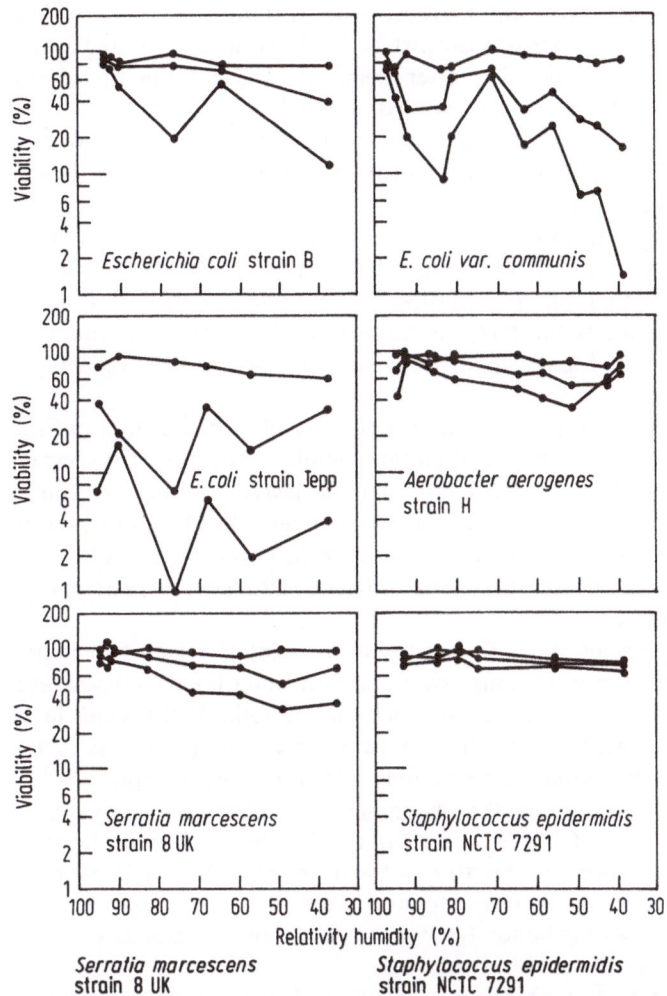

Fig. 14. The effect of relative humidity and aerosol cloud age upon the survival of various bacteria. The plots represent (reading downwards) survival of populations recovered from aerosols into impingers at 1.2 s, 5 and 30 min after generation. (Figure after Anderson et al. [60])

its temperature. The change in temperature during the process of atomization will cause a change in the vapour pressure of the droplet surface.

The chilling of cells of *E. coli* to 4 °C caused rapid death [148]. A large temperature drop could produce a similar response in an aerosol. Consequently, Webb [80] states that survival in air will be greater if the evaporation process is carried out at a higher temperature (i.e., the faster the evaporation rate the better the chance for survival). However, as noted previously, within a certain range, higher survival is attained at lower temperatures. Therefore, at a lower temperature, the temperature drop caused by aerosolization is minimized and the evaporation rate is slower than at higher temperatures.

RH could, at constant temperature and pressure, exert its influence by controlling the rate of evaporation from an aerosol droplet [96, 50]. Even in a saturated water vapour atmosphere, the aerosol droplet, after being sheared off the tip of the atomizer by the airstream of the atomizer, will lose water [155].

The rate of evaporation can be determined by the particle size, the nature and composition of the dissolved substances in the droplet and the diffusion coefficient of the evaporating material.

When the aerosol droplet is transferred from the liquid in the atomizer to a lower RH (air), evaporation occurs until a balance between vapour pressures is attained. Kethley et al. [155] estimate that this equilibrium will be attained in 0.01 to 0.1 s depending on the RH; the lower the RH, the more rapid is the evaporation. This could be an important reason for the high death rate reported in the first second of aerosolization.

A rapid loss of water from cell particles in air could result in one or more of the following: (i) weakening and/or collapse of certain cellular structures; (ii) concentration of droplet or intracellular constituents to toxic levels resulting in cellular poisoning or osmotic lysis; or (iii) inactivation of cellular functions by an imbalance of water throughout the cells [43]. With aerosols of E. coli, Poon [96] concluded that the rate of death and the initial rate of evaporation followed a similar pattern with respect to temperature and RH.

After free water in a droplet starts to evaporate, a point is reached at which cellular water begins to evaporate. This may cause protein-water bonds to break and may possibly start the inactivation process for airborne bacteria. If the water layer surrounding the microbe is evaporated within a very short time, the bonds of the water molecules will be broken almost immediately after the aerosol is sprayed [2].

Evaporation seems to coincide with the phenomenon of increased survival with increasing RH. Hayakawa and Poon [2] found the death rate of airborne bacteria to be slower than the evaporation rate because of the different boundary thicknesses the water molecule has to cross in order to evaporate. The loss of water may result in the loss of protein-water bonds (protein stability) or the exposure of cell components to the oxidizing action of oxygen.

Cox [40, 93] demonstrated that, with aerosols of E. coli, survival was greater at low rather than high RH in inert atmospheres. He concluded that the initial evaporation rates of the aerosol droplets did not influence the long term survival. Nevertheless, evaporation will proceed faster at low RH and this would seem to support the hypothesis of Webb [80] that the faster the evaporation, the greater is the chance for survival.

Decreased recovery of Flavobacterium sp. from the aerosol at extremely high RH could be due, in part, to the equilibration of the airborne particles to a diameter larger than 5 µm [34]. However, equilibration of cells with the environment has been discounted by Cox [157, 29] as the cause of differences between survival of wet and dry disseminated aerosols. The importance of the evaporation rate of airborne cell-containing particles has not been elucidated clearly but it would appear to play a part in the survival of microbial aerosols.

6.4 Cell Water Content and Osmotic Pressure

Water accounts for approximately 77% of vegetative cells [158]. Consequently, biological material can be said to be hydroscopic. A viable cell will always contain some moisture in the airborne state.

The water within the cell, according to the ion resin model [159], seems neither to be as tightly bound as, either clathrate hydrates [160, 161], or hexagonal ice [162] nor as free to orient as liquid water. Cell water is somewhat situated between these extremes.

Different studies suggest that the water content of bacteria is related to the RH of the environment [163, 22, 5, 164]. The rate of transfer of water vapour between the cell surface and the environment is also affected by the surrounding RH [5].

Changes in the water content of lyophilized *S. marcescens* as a function of the aqueous vapour pressure was described by Bateman et al. [164]. Approximately 85% of the original 'sorbed' water [150] was lost upon equilibration at 40% RH (20 °C) but 10% or less was lost at 97% RH. Hess [37] noted the major changes in water content of aerosolized *S. marcescens* (wet disseminated) to occur between 40 and 97% RH. Maximum aerosol stability was achieved at 40% RH- the maximum water loss value for lyophilized *S. marcescens*. It appears that the greater the water loss, the more stable is an aerosolized microbe. Cox and Goldberg [45] state that the control of aerosol survival appears to be through the water content of *P. tularensis* at the moment of aerosol generation rather than in the aerosol phase. This conclusion is based on their experimental results showing that freeze-drying and reconstituting the bacteria with distilled water before aerosol formation had little effect upon survival as a function of RH.

Freeze-dried bacteria show a hysteresis-like retention of water that suggests that the stability of the bacteria may depend on the history of the particle and/or the direction of water transfer. Bateman et al. [164] explained that the hysteresis could result from processes of slow irreversible changes in the dehydrated organisms or from the existence of a potential energy barrier (e.g. super saturation, failure to dissolve, or the formation of crosslinked structures which do not expand reversibly on wetting). Aerosol survival, because of the similarities with freeze-drying viability, could rely on these processes.

Cox [29], in a study of wet and dry dissemination of *P. tularensis*, found differences in the oxygen toxicity for the two methods of aerosolization. This suggests that the bacterial water content and activity do not control aerosol survival.

The diffusion rates of intracellular molecules and polymeric precursors is dependent upon RH in a zone(s) of water content similar to freeze-dried microbes, as previously implied. At RH values high enough to distend the hydrophilic cell structures to their maximum extent, survival with increasing RH should be considered in terms of the osmotic coefficient Φ [164]. In aerosols, the osmotic pressure inside microbes is at an optimum level in a saturated atmosphere [6]. Therefore, no net water should enter or leave. However, the loss of water from cell solution could result in the irreversible formation of hydrates which in turn could change the osmotic pressure.

Cox [6] states that as the RH decreases, the concentration of solutes in the evaporating droplets surrounding airborne bacteria increases and the osmotic lysis could occur from the loss of cell water.

6.5 The Direction of Water Flow

The layer of non-living material surrounding airborne microorganisms is termed the immediate environment [165]. Puck [166] suggested that for a particle, in an atmosphere containing vapours which are completely miscible with water, condensation and solution will occur. The loss of intracellular water can result in an increase in the osmotic pressure. The cell can gain water in a high RH environment as a result of a decrease in the osmotic pressure as vapour is adsorbed. The concentration of materials which will be achieved in an airborne particle will depend upon the vapour pressure relationships and the nature and composition of the bacterial particle [155]. A change in cellular water (by rehydration or dehydration) can cause a physical disruption of cellular components or create an imbalance in metabolic activity [139] causing the death of the cell. The formation of hydrates within the cell structure depends on the direction of water flow across the cells' barriers [49].

Minimum survival for wet dissemination of *P. tularensis* is between 30 and 60% RH [45]. Apparently, water leaves the bacteria as they come into equilibrium with the RH of their environment. For dry dissemination of these organisms, minimum survival occurred at 80% RH, i.e., these bacteria gain water as they come into equilibrium with their environmental RH [45]. Consequently, the direction of water flow is important with respect to the aerosol survival for differing RH ranges.

6.6 Dehydration

The cause of death of aerosolized bacteria has been attributed to the effects of dehydration [177,99]. The primary lethal effect could be intramolecular rearrangement, concentration of toxic products, or an imbalance of metabolic functions [39].

Observations on the variation of airborne bacterial death rate with variations in the RH led Dunklin and Puck [81] to suggest that the degree of cell dehydration is directly related to the sensitivity of aerosolized bacteria. Webb [80] attributed the lethal effect of dehydration of a structural change in an essential macromolecule when bound water is removed in the aerosol state.

Neither strong salt solutions nor a high concentration of glycerol caused a larger death rate than that evinced in the airborne state [155]. However, comparisons of the kinetics of viability loss of aerosolized and freeze-dried bacteria and viruses show a similarity in their results [36]. Dehydration, therefore, may not be the sole cause of death of aerosolized bacteria but just a component of the overall death mechanism.

6.7 Rehydration

Rehydration of bacteria seems to be involved in the RH influence on aerosol survival [6,7,51,40,93,98,157,29,111,31,52,167,145,43,45]. Rehydration has been associated in the death mechanisms of dried *E. coli* aerosolized into a RH above 70% [111,52].

In nitrogen atmospheres, the death mechanism of *E. coli* seems to be influenced by the way water reenters the bacteria during (Cox [98]) and after collection [40]. Cox [6, 7] attributed rehydration as the cause of death of *E. coli* B and *E. commune* sprayed from raffinose solutions into nitrogen at high RH. The same mechanism could be involved at high RH in atmospheres of argon and helium, which were not found to be completely biologically inert [40].

Lysis on rehydration has been found not to be the primary death mechanism of microorganisms in aerosols [29]. The hysteresis effect can explain the differences in survival vs. RH results for wet and dry disseminated bacteria [49]. However, one component involved in the differences encountered between survival of wet and dry disseminated aerosols is assumed to be rehydration [157, 29].

The process of dehydration or rehydration in an aerosol seems to be involved with the formation of hydrates and the bonding of water in the cell.

6.8 Hydration and Adsorption

The surface tension of moist particles governs the mole fraction of adsorbed vapours. Lowering of the surface tension of water by certain chemicals results in selective adsorption of these substances onto the surface and an increase in the mass of the particle. An example of some of these substances are proteins and carbohydrates. Water soluble vapourized organic compounds with a surface tension less than the bacterial droplet surface tension would also be selectively disolved [1].

The response to relative humidity shown by airborne microbes could be a result of the formation of hydrates of key biological molecules [169, 170, 168]. The hydrates could be formed in a semireversible or functionally irreversible manner. Couper et al. [171] have shown that enzyme inactivation can occur at certain RH values. The inactivation could be a result of the formation of different hydrates, some of which may be toxic or irreversible.

Cox [6] hypothesized that a biologically inactive moiety may be produced from hydration of a DNA molecule in a semi-reversible manner by a process of aerosolization. However, Benbough [52] has concluded that DNA synthesis is not impaired by aerosolization of *E. coli*. Phage T7 in aerosols of *E. coli* collected at high RH were able to reproduce [111], which rules out impairment of DNA synthesis. Alterations in RNA synthesis did not seem to be adequate enough to drastically suppress colony formation and T7 production [40].

Cox et al. [36] and Cox [49] have derived viability equations that are based on the hydration of subpopulations of a hypothetical molecular species. This biochemical moiety is proposed to form a series of hydrates when exposed to an environment of low water activity. Protein (enzyme) hydration could be the site of the RH effect on survival.

6.9 Water Bonds and Proteins

Damadian [159] has proposed the theory that cell water orientation is important in the transport system of microorganisms. Nuclear magnetic resonance studies have demonstrated that cell water is more structured than elemental water and less structured than ice [172]. Structured water advocates maintain that structured cell water interacts with charged macromolecules inside cells.

Damadian [100] proposes that a cellular resin is present in the cell and is composed of a fabric of charge groups of the structural polymers of the cell (nucleic acids, proteins, and phospholipids). Water molecules act as dipoles and line up against the charged groups (side chain amino and carboxyl groups and phospholipid and nucleic acid phosphates) of these structural polymers. The result is concentric layers of oriented water molecules about a macromolecule. As the distance from a charged macromolecule increases, order in the water layer decreases. Water structuring, because of its dielectric constant (which influences the force of attraction between two ions), regulates the selection of one counterion over another [173]

The removal of the most firmly held water molecules results in some loss of bacterial stability [163]. Ferry et al. [95] suggest that it is the rate of transmission of water through the cell boundary that is involved in death in bacterial aerosols. This movement of water molecules in and out of the cells in an equilibrium system could result in a collapse of the natural structures of the cell protein. For example, cells of *E. coli* that are freely permeable to ions, small molecules, sugars and peptides are air sensitive. Cells of *Staphylococcus citreus* and *S. aureus*, that are not so permeable, are aerosol stable [4]. Consequently, it can be stated that, if water is semi-structured with a wall or membrane component, the more loosely knit or permeable the cell barriers is, the easier it is to lose or gain cell water and the more unstable the cell is in the aerosol.

With air-dried organisms in aerosols below 80% RH, death due to toxic chemicals, irradiation or dessication can be a result of removal or reorientation of bound water [174, 175].

The L-amino acids that compose peptide chains are the primary structures of protein molecules [176]. Protein groups have differences in their affinity for water which would allow for a large variation, from weak to strong, in the strengths of the protein-water bonds [80].

In an airborne cell, bonded water exchange with atmospheric water vapour could be determined by the strength of protein-water bonds. The many weak bonds could cause cell death but it might take breakage of only a few strong bonds to cause the same result [80]. An increased exchange of water molecules with the environment increases the possibility of a hypothetical irreversible change in the protein structure as a result of the removal of bound water. Webb [80] postulated that a dynamic system exists in the aerosol similar to the one shown below:

$$\text{PROTEIN} \cdot [H_2O]_n + H_2O(g) \rightarrow \text{PROTEIN} \cdot [H_2O]_{n+x} + \text{ENERGY}$$
$$\rightarrow \text{PROTEIN} \cdot [H_2O]_p + (H_2O]_j$$

If the rates of water exchange between water vapour and protein bonded water molecules governs the rate of aerosol death, the relative strengths of the protein-water bonds can be indicated by the inactivation energy for death [80]. Small values of E would suggest that bonds of low energy level are involved in the death mechanism. Ionic, hydrogen, or phosphate bonds may take part in this interaction.

Webb [80] found that, as the cell ages in air, more energy is required to produce its death (i.e. increasing activation energy (E) for death with increasing aerosol age).

If energy is available, the death rate during the latter stages of aerosolization increases more rapidly than that during the earlier stages.

The difference between the E values for inactivation of wet and dry cells at low temperature could be due to the relative amounts of water in the cells. If there were several water layers covering one site of the protein water bond, an 'attack' on this vital site would be more difficult than if no protective layer existed [80].

Bound water appears to occupy strategic positions in protein molecules and only certain chemicals forming hydrogen bonds analogous to protein-water bonds can replace them and maintain the biological integrity of the macromolecule [22,177]. The position of the bond in the molecule also governs the lethality of a protein-water bond rupture or formation.

Cox and Baldwin [178] found that the gaseous atmosphere is involved in the death processes which occur at high RH, independent of oxygen. Gas hydrates which are able to modify a water lattice can be formed by nitrogen and argon. These two gases and helium can confer stability to a more polarizable solute clathrate that is present [179]. Therefore, the slight lethality of atmospheres of nitrogen, argon and helium may act through the modification of cell water structure.

If a biochemical moiety $(A \cdot nH_2O)$ is present in airborne bacteria, most of the water in this solution will evaporate between 60 and 100 % RH [36]. Below 60 % RH, sorption isotherms for bacteria show that water losses are small [180,164]. An environment conducive to the existence of species $A \cdot nH_2O$ has been hypothesized to occur in the interspace between the cytoplasmic membrane and the cell wall [36] where the toxic action of oxygen is thought to occur [111,40,93,53].

Detachment of bacterial components involved in the transfer of substances through the bacterial membrane has been shown for two strains of aerosolized E. coli. Incubation of the bacteria, after collection, with leaked components, partially restored transport activity of methyl-(α-D-gluco)pyranoside [62]. The leaked components were not identified but they might be counterions mentioned in the cellular resin model [173] or constituents of an active transport mechanism.

7 The Stability of Mycoplasmas, L-Forms and Algae in Air

Bacteria and mycoplasmas differ greatly in cell size, composition and morphology of the cell boundary. Nevertheless, the uniqueness of these microbes does not afford them any less stability in air within a certain relative humidity (RH) range [41,43] than most other bacterial species [50]. Airborne M. pneumoniae equilibrated at a dry or wet humidity range was shown to be highly sensitive to the effects of atmospheric moisture immediately after an abrupt change to a mid-range humidity [43]. A sudden change in RH which occurs in natural atmospheres can modify the biological stability of these microorganisms. The survival of this organism was also found to be a function of both RH and temperature but the effects of temperature could only be observed if some water vapour were present [32].

Aerosolized streptococcal L-forms respond to RH in a manner similar to mycoplasma [41,43,42]. Since mid-range RH values were also the most lethal for certain airborne bacteria [81,144,60], it seems that the microbial death mechanisms involved in bacterial aerosols is independent of unique morphological features associated with the cell wall.

Atomized algae (*Nannochloris atomus* and *Synechococcus* sp.) evinced the greatest loss in viability during the first minute after atomization. This initial inactivation is similar to other aerosolized microorganisms [181,145]. The subsequent long term survival was a function of RH [48].

The blue green algae *Synechococcus* sp. showed a pattern of survival resembling that of bacteria in that it remained viable throughout the RH spectrum. The eukaryote, *N. atomus*, remained viable only at near saturated humidities [48].

8 The Effects of Radiation on Airborne Microbes

Disinfection of airborne microbes by ultraviolet (UV) irradiation is dependent upon certain conditions including type of microorganism, suspending menstruum, atmospheric humidity, volume of space, quality of the irradiation, strength and length of the UV ray, total exposure, uniformity of exposure, and air motion.

Ultraviolet irradiation killing of airborne microorganisms has not been found to depend on any RH effect [183,184]. However, Gates [184] and Riley and Kaufman [44]

Table 6. The effect of simulated solar radiation, type of collection medium (agar) aerosol age, and tracer on the aerosol survival of *Serratia marcescens*. (Data after Dorsey et al. [55])

Sampling time (min)	Light intensity (mcal/ cm² min⁻¹)	Per cent recovery with indicated medium[a]			
		Blood agar base		Casitone agar	
		Uranine	No uranine	Uranine	No uranine
2	0	48.9 (46)[b]	53.9 (40)	46.4 (40)	49.6 (37)
4		48.8 (41)	56.6 (39)	45.4 (37)	51.9 (39)
8		44.5 (40)	49.2 (41)	41.2 (37)	44.6 (36)
16		34.9 (45)	41.8 (36)	30.5 (44)	36.8 (37)
32		24.9 (54)	33.2 (35)	22.2 (52)	29.6 (35)
64		14.1 (51)	16.9 (23)	12.0 (50)	14.9 (24)
2	32	45.7 (27)	41.7 (41)	44.4 (23)	40.6 (36)
4		40.8 (32)	42.2 (29)	38.6 (31)	38.3 (33)
8		34.0 (35)	30.9 (31)	31.3 (40)	26.9 (30)
16		18.5 (30)	21.3 (26)	15.5 (26)	17.2 (18)
32		3.6 (44)	8.4 (35)	3.1 (49)	6.9 (36)
64		0.05 (89)	0.4 (36)	0.03 (116)	0.4 (50)
2	70	32.6 (35)	49.6 (41)	30.5 (36)	46.9 (44)
4		27.4 (42)	37.5 (26)	24.4 (47)	34.3 (22)
8		18.1 (48)	25.5 (23)	15.4 (47)	23.5 (20)
16		3.6 (41)	11.6 (35)	2.5 (46)	10.0 (41)
32		0.1 (80)	1.0 (44)	0.08 (89)	0.9 (45)
64		0.00006 (160)	0.009 (63)	0.00003 (175)	0.007 (69)

[a] Mean of six runs for each condition
[b] Figures in parentheses are coefficients of variation rounded off to the nearest whole number

demonstrated a loss of UV sensitivity above RH values of 60%. Response variations between organisms has also occurred. For example, aerosols of *S. marcescens* in the middle RH range were found to be more sensitive to inactivation [44] than other organisms [24].

Kundsin [26] found that droplet nuclei of bacteria and mycoplasmas were destroyed instantly by UV irradiation with very few survivors. The same study demonstrated that L-forms and *Candida albicans* retained 11 to 31% of their droplet nuclei after irradiation. Protection was hypothesized to occur through the action of budding or clumping.

Simulated solar (xenon) radiation caused a significant decrease in viability with aerosols of *E. coli* or *S. marcescens*; Table 6 [55]. Their decay curves showed at least two mechanisms of inactivation, one due to aerosolization, the other to irradiation. When sodium fluorescein was used for the assay method, an additional adverse effect was found. The adsorption of the radiation by nucleic acids can form thymine polymers which could be additive to the lethal effects of aerosolization.

9 Protecting Agents in Aerosols

Hambleton [94] found some substances that improved the survival of bacteria in an aerosol. The mechanism(s) of protection could be the limitation of water loss from the microbe [6], the replacement of structural water [22], or the interaction of the substance with specific intracellular sites [52]. Cox [6] hypothesized that good protective agents for *E. coli* aerosols must prevent or moderate both the toxic action of air and the rate of viability loss within critical relative humidity (RH) ranges.

Polyhydroxy compounds have been used as protective agents for aerosolized bacteria [22] because they form supersaturated viscous layers around bacteria and limit the loss of water [6] in the same way as those bacteria that possess a high lipopolysaccharide content [52]. Micromolar quantities of paramagnetic molecules such as Mn^{+2}, NO_2^-, I^-, and compounds such as ascorbic acid, aminothiols, and reduced dyes have been used and afford protection through a reaction or stabilizing action on the oxygen inactivating site [52]. Protective agents (spray additives) differ in their ability to penetrate the cell wall [6, 51]. Zimmerman [185] suggested that nonpermeable sugars protect airborne microorganisms through a plasmolytic dehydration of the microbes. Hambleton [94] stated that raffinose and dextran do not penetrate the cell wall, glucose and sodium glutamate (which does not permeate the bacterial membrane) entered very slowly, and glycerol was capable of penetrating the bacteria rapidly. Effective concentrations of agents that form supersaturated solutions around bacteria [6], aid survival of aerosolized microbes by preventing changes occurring in the bacterial wall and membrane and by protecting them from the lethal effects of extracellular enzymes. Protective agents do not operate through a modification of the initial evaporation rate of the aerosol particle [6, 40].

Slight protection of *E. coli* K 12 Hfr1 aerosols at high RH was found for glycerol and raffinose, at low RH glycerol was toxic but raffinose was highly protective [93] The positioning of raffinose outside the cell wall was able to confer stability in a similar manner to that for *E. coli* Jepp [51]. Cox [49] states that the protective action of raffinose is related to its stabilizing action upon hydrates of a hypothetical active biochemical moiety.

Cox [29] found that peptone broth spraying fluid either interacts with oxygen to yield a toxic product or does not completely protect against oxygen-induced death since survival in air for *Pasteurella tularensis* was less than in nitrogen. When a 1% peptone diluent was used in the viability assay medium, the survival of *P. pestis* aerosolized into a high RH (65 and 87%) was adversely affected. However, heart infusion broth, used as diluent, increased the number of viable cells in aerosols held at high RH values [38].

S. marcescens recovery from the airborne state showed no change in either minimal or nutrient settling plate agar mediums. Plain gelatin liquid impingement medium, however, showed evidence of enhanced recovery with the highest recovery resulting from the enrichment of the medium [186].

10 *Serratia marcescens*

Serratia marcescens has been used by Orr and Gordon [156], Dimmick [5], and Cox et al. [36] in aerosol viability studies. The microbe is relatively nonpathogenic and easily grown in batch conditions. The distinctive red pigmentation allows for quick identification on agar plates in the presence of contaminating microbes.

S. marcescens is a facultative anaerobe widely distributed in water, soil, and food. It is occasionally found in pathological specimens and has been identified in causes of outbreaks of nosocomial infections [187, 188]. Bacteriocins (marcescins) produced by this microbe are proteinaceous antibiotics that could play a role in noscomial infections [189].

The synthesis of a prodigiosene (prodigiosin, Fig. 15) causes a red pigmentation in colonies. Maximum prodigiosin production has been attained with 96 h (senescent) cultures at 27 °C [190]. Prodigiosin synthesis is a variable characteristic (as shown by the appearance of white colonies) influenced by culture conditions. Nonproliferating (stationary phase) cells synthesize this pigment under alkaline conditions with amino acids and acetate as the substrates. Inhibition of prodigiosin production is

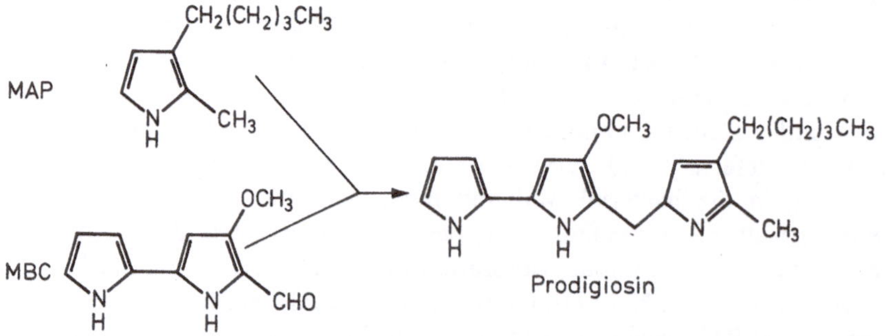

Fig. 15. The structure of prodigiosin and its precursors. MAP denotes 2-methyl-3-amylpyrrole; MBC denotes 4-methoxy-2,2'-bipyrrole-5-carboxyaldehyde. (Figure after Williams [190])

brought about by the addition of glucose, phosphate, or chloramphenicol to the medium.

Prodigiosin has been categorized as a secondary metabolite of *S. marcescens* [190]. The monopyrrole precursor (2-methyl-3-amylpyrrole) has been shown to require at least four enzymes for its formation and five enzymes have been identified in the biosynthesis of the bipyrrole (4-methoxy-2-2′bipyrrole-5-carboxy-aldehyde); a tenth enzyme is involved in the coupling of these two components to form prodigiosin. The pigment is located in the cell envelope [190].

Aerosolization of *S. marcescens* into a standard test condition (20 °C and 65% RH) showed a variation in response that was not completely explained by the appearance of coloured variants (differing prodigiosin or colour concentration) from the blood-red wild type parent strain [186].

However, the reduced airborne viability was correlated to the emergence of a wide variation in the morphology and dimensions of the cells. An isolated variant culture (white, pink or red in appearance) was noted to decrease in airborne stability with an increase in the variety of cellular forms and sizes [186].

Cell envelope glycoproteins, extracted by sodium dodecyl sulphate, from isolated walls of two strains of *S. marcescens* were found to be very similar in chemical composition [191]. Both fractions contained approximately 50% proteins and 10% carbohydrates. The only sugar components identified in the carbohydrate moiety were glucose and glucosamine (refer to Fig. 4 for peptidoglycan structure).

Orr and Gordon [156], using the Millikan experiment, calculated the density (w/v) of aerosolized *S. marcescens* cells. The values of the density were found to be similar, over RH values from 40 to 100%, to liquid-phase sedimentation studies which indicated a density from 1.057 to 1.062 g cm^{-3}. At RH values below 40%, calculations showed a sharp decrease in the apparent density to a minimum of approximately 0.913 g cm^{-3} at the lowest RH studied. The decrease in density was attributed to the inclusion of air into spaces produced by a wrinkling of the cell wall at low RH values.

The average diameter of *S. marcescens* cells, aerosolized from beef extract broth at 20% RH, was found to be 0.62 μm ($\pm$0.02 standard error) when measured by an electron microscope after collection on an oil film [155]. Orr and Gordon [156] measured the radius of *S. marcescens* cells at the same RH but aerosolized from distilled deionized water by means of electron photomicrographs and found a mean radius of 0.4 μm (or 0.8 μm diameter). Evidently, the diameter of airborne cells depends on the spraying medium.

Above RH values of 75%, aerosols of *S. marcescens* have been found to be relatively stable with respect to viability in air [186], in CO [46], and after exposure to UV irradiation [44]. Below this RH value, viability variations in response to the aerosol state generally increased.

Electron paramagnetic studies have shown that freeze-dried *S. marcescens* cells contain two unidentified molecular species that can react with oxygen [36]. Since the site of oxygen toxicity is believed to lie in the interspace between the cytoplasmic membrane and the cell wall [51], these components are assumed to be located within this area.

There are many stresses acting on the viability of aerosolized microbes which may act collectively or individually depending on the environmental conditions. If, indeed, there is a component of an airborne microorganism that is affected by aerosolization, as evidenced by the viability response of the cell, this component(s) must, at least, be located in the cell wall, have a site for oxidation, be neutralized in its toxic effect by chloramphenicol [31], and be susceptible to hydration or the absorbed water effect.

11 Review of Electrostatic Forces Relevant to Electrodeposition of Bacterial Aerosols

Millikan in 1911 [192] devised an experimental procedure to measure the charge unit on an oil drop in an electric field. In the absence of an electric field, a droplet falling through still air is acted upon by two forces (Fig. 16):
a gravitational force

$$F_g = m_p g \tag{7}$$

where m_p = mass of the particle, gm
$\quad\ g$ = gravitational constant, cm s^{-2}
and an opposing frictional force of air drag

$$F_d = 6\pi\eta r_p v_p \tag{8}$$

where η = air viscosity, centipoise
$\quad r_p$ = radius of the particle, cm
$\quad v_p$ = speed of the particle, cm s^{-1}.
When the speed of the particle is constant ($F_g = F_d$) these forces balance at some downward velocity and the terminal velocity can be described by Stokes Law:

$$v_t = \frac{P_p d^2 p g}{18\eta} \tag{9}$$

where v_t = terminal velocity, cm s^{-1}
$\quad \varrho_p$ = particle density, gm cm^{-3}
$\quad d_p$ = particle diameter, cm.

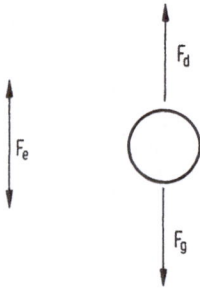

Fig. 16. Forces that act on a falling particle in an electric field. F_g is the gravitational force downward; F_d is the drag force upward; F_e is the electrical force in either direction depending on the polarity of the particle, sign of the electric field, and surface charge on the particle

If the particle is aerosolized into a uniform electrical field, the total force acting on the particle is a combination of Eqs. (1) and (2) and the electrical force:

$$F_{t_{down}} = F_g + F_{el} - F_d$$
$$= m_p g + q_p \varepsilon - 6\pi r_p v_p \eta \qquad (10)$$

where F_{el} = the electrical force exerted in the downward direction, dyne.

With no acceleration, the terminal velocity can be expressed as:

$$v_t = \frac{V_p g \varrho_p + q_p \varepsilon}{6\pi r_p \eta} \qquad (11)$$

where V_p = volume of the particle, cm^{-3}.

However, aerosol particles can accumulate charge in the process of being aerosolized (e.g. contact electrification). The aerosol particle density in a given space decreases as a result of charge repulsion which causes the particles to migrate to the boundaries of the enclosed space. This decrease in particle numbers can be expressed as [1]:

$$\frac{n}{n_0} = \frac{1}{1 + \frac{n_0}{E} q_p^2 K_p t} \qquad (12)$$

where n = cell concentration at time t, particles $\cdot$ cm^{-3}
  n_0 = initial cell concentration, particles $\cdot$ cm^{-3}
  t = time, s
  q_p = charge on particle, coul
  ε = permittivity, $coul^2$ $dyne^{-1}$ cm^{-2}
  K_p = particle mobility, cm V^{-1} s^{-1}.

Cunningham's correction factor accounts for the discontinuous nature of air on a small size scale (near the mean free path of molecules in air). For very small particles, <0.5–0.1 μm, there is a noticeable error in the Stokes velocity due, apparently, to momentary, statistical fluctuations in air density.

If the particles are collected on agar in a plastic Petri plate, preferential deposition on the agar surface can result from the accumulation of electrostatic charge that causes a localized field on the exposed rim of the Petri plate. The plastic is an insulator and, in an electric field, built-up charge cannot bleed off.

A particle of 1 μm radius and 100 unit charges in a uniform electric field of 2000 V m^{-1} would obtain a maximum velocity of 2 cm s^{-1} [1]. However, distortion of the field would cause angular displacement and acceleration. The overall force on the particle would have to be calculated with the above variables taken into consideration.

12 Cell Wall Charge

Aerosolized *E. coli* (MRE strains 160 and 162) show alterations in surface properties as demonstrated by the change in the electrophoretic mobility of the bacteria [94, 62]. At a particle's surface, a continuous exchange of water molecules in equilibrium with the H_2O vapour of the air occurs. At RH's above 50%, charge dissipates rapidly; at RH's below 30%, the charge may remain long enough to be significant [1]. The charge of an aerosol particle increases with size and possibly depends on the environmental air temperature [69].

Structures containing flexible amphoteric polyelectrolytes (e.g. peptidoglycans) cannot be assigned a value for charge density because of the possible extensive internal neutralization of charge [125]. Electrostatic interactions among the fixed ionized groups in wall polymers of *Staphylococcus aureus* and *Micrococcus lysodeikticus* caused the contraction and expansion of isolated cell walls in response to changes in the environmental pH and ionic strength. Whole cells of *Bacillus megaterium* were found to undergo changes in volume resulting from changes in electrostatic interactions of cell wall polymers, especially amphoteric peptidoglycans [193].

M. lysodeikticus cell walls suspended in medium of high pH (11–12) and relatively low ionic strength (0.02) were expanded polyanions (most of the amino groups were neutral and all of the carboxyl and phosphate groups were negative). Lowering the environmental pH to ca. 4.5 caused protonation of the amino groups, the electrostatic attraction of the anionic carboxyl groups, and a maximum contraction of the cell walls. This pH value is close to the predicted isoelectric pH

Table 7. The approximation of surface charge densities of microorganisms

Microorganism	Surface charge density ($Coulomb \times cm^{-2}$)
Escherichia coli	-1.63×10^{-6}[a]
	-1.33×10^{-6}[b]
Streptococcus faecalis	-6.94×10^{-7}[a]
Typhoid bacilli	-1.13×10^{-6}[c]
Aerobacter aerogenes	-9.67×10^{-7}[d]
	-9.27×10^{-5}
	-7.81×10^{-5}
	-9.27×10^{-7}[f]
	1.05×10^{-4}
	7.17×10^{-5}[e]
Streptococcus pyogenes	-1.13×10^{-6}[gh]

[a] Schott, H. and Young, C. Y. [196]
[b] Haydon, D. A. [197]
[c] Abramson, H. A. [198]
[d] Plummer, D. T. and James, A. M. [194]
[e] Gittens, G. J. and James, A. M. [137]
[f] Gittens, G. J. and James, A. M. [138]
[g] Measure of the surface Carboxyl groups
[h] Hill, M. J., James, A. M., and Maxted, W. R. [139]

for cell walls (pI ~ 4.3). At lower pH, the carboxyl groups were protonated and the cell walls were converted to polycations. Electrostatic repulsion caused expansion of this structure [125].

Plummer and James [194] found that the only ionogenic groups on the surface of *Aerobacter aerogenes* were carboxyls. If the area of a carboxyl group is approximately 20 A², Gittens and James [137, 138] state that successive layers of charge must contribute to the electrokinetic charge for recorded surface charge densities of 2.15×10^5 esu $\cdot$ cm^{-2} (diazomethane treated surface). Individual amino acid carboxyl groups contributing to the surface charge can be found by the methods of Ingram and Salton [140] and Hill et al. [139].

Electrophoretic investigations on surface charge have inherent difficulties associated with their measurement e.g. dielectric constants of the materials, conductivity of the medium, and relative ionic strengths of comparable mediums). Statistical approaches to the orientation of water in the membrane and its configurational influence on the membrane components that constitute the charge of the microbial surface are not very conclusive. However, some investigators have used microelectrophoresis or other methods to obtain a quantiative surface charge density (Table 7). The majority of the surface charge densities are negative but the variation is over two orders of magnitude. The total charge density is going to be modified by the amount of gegenion adsorption.

Riddick [195] correlates the zeta potential to ion adsorption and the flocculating ability of microbes. The zeta potential can be calculated from:

$$Z = \frac{4\pi\delta q}{D^*} = \frac{4\pi\eta u}{D^*}$$

where Z = zeta potential, mV
δ = diffuse counterion layer thickness, cm
q = charge on the surface of the particle, coul
D^* = dielectric sonstant of the suspending fluid
η = viscosity of the suspending fluid, poise
u = electrophoretic mobility, cm² s^{-1} V^{-1}.

A large electrophoretic mobility would indicate that an aerosol cloud would contain particles that would have an increased repulsive force. This would cause the particles to be dispersed to the chamber walls faster than if the particles were neutral (i.e. zero electrophoretic mobility). These aerosol droplets would also show a greater terminal velocity in an electric field.

An electrostatic sampler employs an electric field to deposit airborne microorganisms onto the surface of agar plates. Luckiesh et al. [67] employed this instrument to collect bacteria onto either a positively, negatively or neutrally charged agar plate. They concluded that both positively and negatively charged bacteria exist simultaneously in the air.

Gregory [3] found that basidiospores of *Gonoderma applanatum*, falling between two condenser plates charged positively and negatively with respect to earth potential, usually carry a net positive charge. The colony-forming units of *E. coli*, atomized from an aqueous suspension into a room, were 10 times as numerous on the positive electrode as on the negative electrode. However, 30% of the naturally occurring airborne bacteria were deposited on the negative electrode [67]. Therefore, the aerosol medium could contribute to the charge on the surface of the aerosolized microbe.

Orr and Gordon [156] used a modified Millikan apparatus and calculated the surface charge of airborne *Serratia marcescens*. The charge ranged from a deficiency of approximately 5 electrons to more than 20 electrons. They also found that the calculated charges were slightly greater than the nearest standard electron charge when RH's above 50% were employed.

An electric charge on the surface of the droplet will be enhanced, by charge repulsion and expansion, the evaporation of the liquid surrounding the bacteria (i.e. the larger the repulsion of molecules, the lower will be the retentive inter-molecular bonds). Assuming there is a net electric charge on an aerosol droplet, as the water evaporates, the electric charge will be retained on the remaining surface area. Consequently, the electric field (charge/unit surface area) will increase. However, if the electric field at the surface increases to a large enough value, ionization by air ions could ultimately reduce the magnitude of the charge.

This discussion leads to the conclusion that microbial surface charge is highly variable in artificially and naturally occurring aerosols and solutions. The ideal situation for depositing the majority of suspended airborne particulates would be to charge the particles to one polarity at the point of aerosolization and then deposit them in an intense electric field.

13 Nomenclature

A		Ångstrom
ADR		apparent dilution ratio
A · n H$_2$O		a biochemical moiety in airborne bacteria
BDR		biological dilution ratio
CFU		colony forming unit
D*		dielectric constant of suspending fluid
DATA		dual aerosol transport apparatus
DHD		dynamic-humidity-death
DNA		deoxyribonucleic acid
dN		change in aerosol particle (or microbial number)
d$_p$	cm	particle diameter
dt		change in time
E		activation energy for death of cells
EDTA		ethylene diaminetetra-acetic acid
e	coul.^{2} dyne^{-1} cm^{-2}	permittivity
F$_d$		fractional force of air drag
F$_e$		electrical force in either direction
F$_{el}$	dyne	electrical force in downward direction
Fg		gravitational force
g		gravitational constant
pI		isoelectric point
k		rate constant

Kp	cm v^{-1} s^{-1}	particle mobility
MCB		4-methoxy-2,2′bipyrrole-5-carboxyaldehyde
MAP		2-methyl-3-amylpyrrole
m$_p$		mass of particle
N (or n)		microbial (particle) number
N$_{BI}$		number of tracer spore CFV in the control sample
N$_{BT}$		number of tracer spore CFV at time t
N$_{VI}$		number of CFV of the test organism in the initial sample
N$_{VT}$		number of CFV of the test organism at time t
q$_p$	coulomb	charge on particle
RH		relative humidity
r$_p$	cm	radius of the particle
RNA		ribonucleic acid
u	cm^2 s^{-1} N^{-1}	electrophoretic mobility
UV		ultraviolet irradiation
v$_p$	cm s^{-1}	speed of a particle (in one example, particle volume)
v$_t$	cm s^{-1}	terminal velocity
Z	mv	zeta potential
δ	cm	diffuse counter-ion thickness
η	centipoise	air viscosity
P$_p$	g cm^{-3}	particle density
Φ		osmotic coefficient

14 References

1. Dimmick, R. L., Akers, A. B., (ed.): Introduction to Experimental Aerobiology, New York: Wiley-Interscience 1969
2. Hayakawa, I., Poon, C. P.: Ind. Hyg. J. *26*, 150 (1965)
3. Gregory, R. H.: Microbiology of the Atmosphere, New York: Interscience Publications, Inc. 1961
4 Goodlow, R. J., Leonard, F. A.: Bacteriol. Rev. *25*, 182 (1961)
5. Dimmick, R. L.: J. Bacteriol. *80*, 289 (1960)
6. Cox, C. S.: J. Gen. Microbiol. *43*, 383 (1966a)
7. Cox, C. S.: J. Gen. Microbiol. *45*, 283 (1966b)
8. Spendlove, I. C.: Dev. Ind. Microbiol. *15*, 20 (1974)
9. Lighthart, B., Frisch, A. S.: Appl. Environ. Microbiol. *31*, 700 (1976)
10. Zobell, C. E.: Marine Microbiology, Chronica Botanica Co., Waltham, Mass., U.S.A., 1946
11. Parker, B., Garsom, G.: Bioscience *20*, 87 (1970)
12. Woodcock, A. H.: J. Hyg. *42*, 339 (1955)
13. Blanchard, D. C., Syzdek, L.: Sci. *170*, 626 (1970)
14. Fulton, J. D.: Appl. Microbiol. *14*, 232 (1966a)
15. Fulton, J. D.: Appl. Microbiol. *14*, 245 (1966c)
16. Fulton, J. D.: Microorganisms of the upper atmosphere. V. Relationship between frontal activity and the micropopulation at altitude, Appl. Microbiol. *14*, 245 (1966c)
17. Southey, R. F. W., Harper, G. J.: J. Appl. Bacteriol. *34*, 547 (1971)
18. Buchbinder, L., Solowey, M., Solotorovsky, M.: In: Comparative studies of bacteria in air in enclosed spaces, Heat. Piping Air lond. 1945, p. 389, 1945

19. Kraidman, J.: Report for the American Association for Contamination Control, 1975
20. Doetsch, R. N., Cook, T. M.: In: Introduction to Bacteria and their Ecobiology, Baltimore, Maryland: U.S.A., University Park Press 1973
21. Imshenetsky, A. A., Lysenko, S. V., Kazakov, G. A.: Appl. Environ. Microbiol. *35*, 1 (1978)
22. Webb, S. J.: Can. J. Microbiol. *6*, 71 (1960)
23. Kundsin, R. B.: J. Bacteriol. *91*, 942 (1966)
24. Wells, W. F.: Airborne Contagion and Air Hygiene, Cambridge, Mass.: Harvard University Press 1955
25. Hatch, T. F.: Im. Rev. Respirat. Diseases *83*, 412 (1961)
26. Kundsin, R. B.: Appl. Microbiol. *16*, 143 (1968)
27. Pelczar, M. J., Reid, R. D.: Microbiology, New York: McGraw-Hill Book Co., 1972
28. Anderson, J. D.: J. Gen. Microbiol. *45*, 303 (1966)
29. Cox, C. S.: Appl. Microbiol. *21*, 482 (1971)
30. Akers, T. G., Bond, S., Goldberg, L. J.: Appl. Microbiol. *14*, 361 (1966)
31. Hatch, M. T., Dimmick, R. L.: Bact. Rev. *30*, 597 (1966)
32. Wright, D. N., Bailey, G. D., Goldberg, L. J.: J. Bacteriol. *99*, 491 (1969)
33. Ehrlich, R., Miller, S., Walker, R. L.: Appl. Microbiol. *19*, 245 (1970a)
34. Ehrlich, R., Miller, S., Walker, R. L.: Appl. Microbiol. *20*, 884 (1970b)
35. Ehrlich, R., Miller, S.: Appl. Microbiol. *25*, 369 (1973)
36. Cox, C. S., Gagen, S. J., Baxter, J.: Can. J. Microbiol. *20*, 1529 (1974)
37. Hess, G. E.: Appl. Microbiol. *13*, 781 (1965)
38. Won, W. D., Ross, H.: Appl. Microbiol. *14*, 742 (1966)
39. Zentner, R. J.: Bacteriol. Rev. *30*, 551 (1966)
40. Cox, C. S.: J. Gen. Microbiol. *50*, 139 (1968a)
41. Wright, D. N., Bailey, G. D., Hatch, M. T.: J. Bacteriol. *95*, 251 (1968a)
42. Stewart, R. H., Wright, D. N.: Appl. Microbiol. *19*, 865 (1970)
43. Hatch, M. T., Wright, D. N., Bailey, G. D.: Appl. Microbiol. *19*, 232 (1970)
44. Riley, R. L., Kaufman, J. E.: Appl. Microbiol. *23*, 1113 (1972)
45. Cox, C. S., Goldberg, L. J.: Appl. Microbiol. *23*, 1 (1972)
46. Lighthart, B.: Appl. Microbiol. *25*, 86 (1973)
47. Turner, A. G., Salmonsen, P. A.: J. Appl. Bact. *36*, 497 (1973)
48. Ehresmann, D. W., Hatch, M. T.: Appl. Environ. Microbiol. *29*, 352 (1975)
49. Cox, C. S.: Appl. Environ. Microbiol. *31*, 836 (1976)
50. Anderson, J. D., Cox, C. S.: Symp. Soc. gen. Microbiol. *17*, 203 (1967)
51. Cox, C. S.: J. Gen. Microbiol. *49*, 109 (1967)
52. Benbough, J. E.: J. Gen. Microbiol. *56*, 241 (1969)
53. Cox, C. S., Bondurant, M. C., Hatch, M. T.: J. Hyg. *69*, 661 (1971)
54. Cox, C. S., Banter, J., Maidment, B. J.: J. Gen. Microbiol. *75*, 179 (1973)
55. Dorsey, E. L., Berendt, R. F., Neff Jr., E. L.: Appl. Microbiol. *20*, 834 (1970)
56. Douett, H. A., May, K. R.: N. Sci. *41*, 579 (1969)
57. Serat, W. F., Kyono, J., Mueller, P. K.: Atmos. Environ. *3*, 303 (1969)
58. Lighthart, B., Hiatt, V. E., Rossano Jr., A. T.: J. Air Pollut. Control Assoc. *21*, 639 (1971)
59. Cox, C. S., Hood, A. M., Banter, J.: Appl. Microbiol. *26*, 640 (1973)
60. Anderson, J. D., Dark, F. A., Peto, S.: J. Gen. Microbiol. *52*, 99 (1968)
61. Hambleton, P.: J. Gen. Microbiol. *69*, 81 (1971)
62. Benbough, J. E. et al.: J. Gen. Microbiol. *72*, 511 (1972)
63. Kethley, T. W., Fincher, E. L., Cown, W. B.: J. Infect. Dis. *100*, 97 (1957)
64. Mercer, T. T., Goddard, R. F., Flores, R. L.: Annals of Allergy *23*, 314 (1965)
65. Larson, E. W., Young, H. W., Walker, J. S.: Appl. Environ. Microbiol. *30*, 150 (1976)
66. Mercer, T. T., Tillery, M. I., Chow, H. Y.: Am. Ind. Hyg. Assoc. J. *29*, 66 (1968)
67. Luckiesh, M., Taylor, A. H., Holladay, L. L.: J. Bacteriol. *52*, 55 (1946)
68. Cown, W. B., Kethley, T. W., Fincher, E. L.: Appl. Microbiol. *5*, 119 (1957)
69. Houwink, E. H., Rolvink, W.: J. Hyg. *55*, 544 (1957)
70. Andersen, A. A.: J. Bacteriol. *76*, 471 (1958)
71. Decker, H. M. et al.: Appl. Microbiol. *6*, 398 (1958)
72. Wolochow, H.: Appl. Microbiol. *6*, 201 (1958)

73. Batchelor, H. W.: In: Aerosol samplers, Adv. in Appl. Microbiol. *2*, 31 (1960)
74. Andersen, A. A., Andersen, M. R.: Appl. Microbiol. *10*, 181 (1962)
75. Lidwell, O. M., Noble, W. C.: J. Appl. Bact. *28*, 280 (1965)
76. Malligo, J. E., Idoine, L. S.: Appl. Microbiol. *12*, 32 (1964)
77. May, K. R.: Appl. Microbiol. *12*, 37 (1964)
78. Knuth, R. H.: Am. Ind. Hyg. Assoc. J. *30*, 379 (1969)
79. Cox, C. S. et al.: Appl. Microbiol. *20*, 927 (1970)
80. Webb, S. J.: Can. J. Microbiol. *5*, 649 (1959)
81. Dunklin, E. W., Puck, T. T.: J. Exptl. Med. *87*, 87 (1948)
82. Henderson, D. W.: J. Hyg. *50*, 53 (1952)
83. Wolfe, E. K.: Bacteriol. Rev. *25*, 194 (1961)
84. Harper, G. J., Morton, J. D.: J. Gen. Microbiol. *7*, 98 (1952)
85. Harper, G. J., Hood, A. M., Morton, J. D.: J. Hyg. *56*, 364 (1958)
86. Harper, J. G.: Some observations on the influence of suspending fluids on the survival of airborne viruses, Proc. Intern. Symp. Aerobiol. 1st Berkeley, Calif. p. 335, 1963
87. Harstad, J. B.: Appl. Microbiol. *13*, 899 (1965)
88. Green, L. H., Green, G. M.: Appl. Microbiol. *16*, 78 (1968)
89. Miller, W. S. et al.: Appl. Microbiol. *9*, 248 (1961)
90. Strange, R. E., Powell, E. O., Pearce, T. W.: J. Gen. Microbiol. *67*, 349 (1971)
91. Strange, R. E., Martin, K. L.: J. Gen. Microbiol. *72*, 127 (1972)
92. Strange, R. E. et al.: J. Gen. Microbiol. *72*, 117 (1972)
93. Cox, C. S.: Gen. Microbiol. *54*, 169 (1968b)
94. Hambleton, P.: J. Gen. Microbiol. *61*, 197 (1970)
95. Ferry, R. M., Brown, W. F., Damon, E. B.: J. Hyg. *56*, 389 (1958)
96. Poon, C. P. C.: Studies on the instantaneous death of *Escherichia coli*, Am. J. Epidem. *84*, 1 (1966)
97. Beard, C. W., Anderson, D. P.: Avian Dis. *11*, 54 (1967)
98. Cox, C. S.: J. Gen. Microbiol. *59*, 77 (1969)
99. Webb, S. J.: J. Gen. Microbiol. *58*, 317 (1969)
100. Damadian, R.: Structured water or pumps?, Letter, Sci. *193*, 528 (1976)
101. Anderson, J. D., Dark, F. A.: Gen. Microbiol. *46*, 95 (1967)
102. Ennis, H. L., Lubin, M.: Biochim. Biophys. Acta *95*, 605 (1965)
103. Tempest, D. W., Dicks, J. W., Hunter, J. R.: J. Gen. Microbiol. *45*, 135 (1966)
104. Anderson, J. D., Crouch, G. T.: J. Gen. Microbiol. *47*, 49 (1967)
105. Hangaard, N., Hess, M. E., Itskovitz, H.: J. Biol. Chem. *227*, 605 (1957)
106. Hollaender, A., Stapleton, G. E., Martin, F. L.: Nature *167*, 103 (1951)
107. Tallentire, A.: Nature *182*, 1024 (1958)
108. Tallentire, A., Dickinson, N. A., Collett, J. H.: J. Pharm. Pharmacol. Suppl. *15*, 180T (1963)
109. Lion, M. B., Bergmann, E. D.: J. Gen. Microbiol. *24*, 291 (1961a)
110. Lion, B. M., Avi-Dor, T.: Israel J. Chem. *1*, 374 (1968)
111. Cox, C. S., Baldwin, F.: J. Gen. Microbiol. *44*, 15 (1966)
112. Dimmick, R. L., Heckley, R. S., Hollis, D. P.: Nature (London) *192*, 776 (1961)
113. Dimmick, R. L., Heckley, R. S., Hollis, D. P.: Nature (London) *192*, 776 (1961)
114. Heckley, R. J., Dimmick, R. L., Windle, J. J.: J. Bacteriol. *85*, 961 (1963)
115. Bateman, J. B. et al.: Appl. Microbiol. *9*, 567 (1961)
116. Staudie, W. C., Riggs, B. C., Haugaard, N.: Am. J. Med. Sci. *207*, 84 (1944)
117. Thomas, J. J., Neptume, E. M., Suddath, H. D.: Biochem. J. *88*, 31 (1963)
118. Barron, E. S. G.: J. Biol. Chem. *113*, 695 (1936)
119. Gershmann, D. et al.: Sci. *119*, 623 (1954)
120. Rogers, L. A.: J. Infect. Dis. *14*, 100 (1914)
121. Naylor, H. B., Smith, F. A.: J. Bacteriol. *52*, 565 (1946)
122. Caughey, W. S.: Ann. N. Y. Acad. Sci. *174*, 148 (1970)
123. Lehninger, A. L.: Biochemistry, New York: Worth Publishers, Inc. 1970
124. Cairns, J., Denhardt, D. T.: J. Mol. Biol. *36*, 335 (1968)
125. Ou, L. T., Marquis, R. E.: J. Bacteriol. *101*, 92 (1969)

126. Knaysi, G.: Elements of Bacterial lytology, Ithaca, N.Y.: Comstock Publishing Co., 1951
127. DePetris, S.: J. Ultrastruct. Res. *19*, 45 (1967)
128. Salton, M. R. J.: Ann. Rev. of Microbiol. *21*, 417 (1967)
129. Weidel, W., Frank, H., Martin, H. H.: J. Gen. Microbiol. *22*, 158 (1960)
130. Noller, E. C., Hartsell, S. E.: J. Bacteriol. *81*, 482 (1961)
131. Bohinski, R. C.: Modern Concepts in Biochemistry, p. 189, Allyn and Bacon: Boston, Mass. U.S.A., 1973
132. Schnaitman, C. A.: J. Bacteriol. *108*, 553 (1971)
133. Aspell, M. A., Eagon, R. G.: Biochem. Biophys. Res. Comm. *22*, 664 (1966)
134. Winshell, E. B., Neu, H. C.: J. Bacteriol. *102*, 537 (1970)
135. Lieve, L.: J. Biol. Chem. *243*, 2373 (1968)
136. Gel'man, N. S., Lukoyanova, M. A., Ostrovsku, D. N.: In: Respiration and Phosphorylation of Bacteria, New York: Plenum Press 1967
137. Gittens, G. J., James, A. M.: Biochim. Biophys. Acta *66*, 237 (1963a)
138. Gittens, G. J., James, A. M.: Biochim. Biophys. Acta *66*, 250 (1963b)
139. Hill, M. J., James, A. M., Maxted, W. R.: Biochim. Biophys. Acta *71*, 740 (1963)
140. Ingram, V. M., Salton, M. R. J.: Biochim. Biophys. Acta *24*, 9 (1957)
141. Wells, W. F., Zapposodi, P.: Sci. *96*, 277 (1948)
142. Shechmeister, I. L., Goldberg, L. J.: J. Infect. Dis. *87*, 116 (1950)
143. Ferry, R. M., Maple, T. G.: J. Infect. Dis. *95*, 142 (1954)
144. Beebe, J. M.: J. Bacteriol. *78*, 18 (1959)
145. Hatch, M. T., Wolochow, H.: In: Bacterial survival consequences of the airborne state in Introduction to Experimental Aerobiology, Dimmick, R. L. (ed.), p. 267. New York, N.Y.: Wiley-Interscience 1969
146. Williamson, A. E., Gotaas, H. B.: Ind. Med. 11, Ind. Hyg. Sec. *3*, 40 (1942)
147. Loosli, C. G. et al.: Proc. Soc. Exptl. Biol. Med. *53*, 205 (1943)
148. DeOme, K. B.: Am. J. Hyg. *40*, 239 (1944)
149. Monk, G. W. et al.: J. Bacteriol. *72*, 368 (1956)
150. Monk, G. W., McCaffrey, P. A.: J. Bacteriol. *73*, 85 (1957)
151. Davis, M. S., Bateman, J. B.: J. Bacteriol. *80*, 580 (1960)
152. McDade, J. H., Hall, L. B.: Am. J. Hyg. *80*, 192 (1964)
153. Wells, W. F., Riley, E. C.: J. Ind. Hyg. Toxicol. *19*, 513 (1937)
154. Ferry, R. M., Brown, W. F., Damon, E. B.: J. Hyg. *56*, 125 (1958)
155. Kethley, T. W., Cown, W. B., Fincher, E. L.: Appl. Microbiol. *5*, 1 (1957)
156. Orr, C., Gordon, M. T.: J. Bacteriol. *71*, 315 (1956)
157. Cox, C. S.: Appl. Microbiol. *19*, 604 (1970)
158. Black, S. H., Gerhardt, P.: J. Bacteriol. *83*, 960 (1962b)
159. Damadian, R.: Sci. *165*, 79 (1969)
160. Davidson, D. W., Wilson, G. J.: Can. J. Chem. *41*, 1424 (1963)
161. Gough, S. R., Walley, E., Davidson, D. W.: Can. J. Chem. *46*, 1673 (1968)
162. Onsager, L., Dupuis, M.: In: The electrical properties of ice, in Pesce, B. (ed.), Electrolytes, New York: Pergamon Press 1962
163. Scott, W. J.: J. Gen. Microbiol. *19*, 624 (1958)
164. Bateman, J. B. et al.: J. Gen. Microbiol. *29*, 207 (1962)
165. Kethley, T. W., Fincher, E. L., Cown, W. B.: Appl. Microbiol. *4*, 237 (1956)
166. Puck, T. T.: J. Exptl. Med. *85*, 729 (1947)
167. Hatch, M. T., Warren, J. C.: Appl. Microbiol. *17*, 685 (1969)
168. Falk, M.: J. Am. Chem. Soc. *86*, 1226 (1964)
169. Falk, M., Hartman, K. A., Lord, R. C.: J. Am. Chem. Soc. *84*, 3843 (1962)
170. Falk, M., Hartman, K. A., Lord, R. C.: J. Am. Chem. Soc. *85*, 391 (1963)
171. Couper, A., Eley, D. D., Hayward, A.: Discuss. Faraday Soc. *20*, 174 (1955)
172. Kolata, G. B.: Research News, Sci. *192*, 1220 (1976)
173. Minkoff, L., Damadian, R.: Biophys. J. *13*, 167 (1973)
174. Webb, S. J., Cormack, D. V., Morrison, H.: Nature *201*, 1103 (1964)
175. Webb, S. J., Dumasia, M. D.: Can. J. Microbiol. *10*, 877 (1964)
176. Fraser, R. D. B., MacRae, T. P.: Nature *183*, 179 (1959)

177. Webb, S. J.: Can. J. Biochem. Physiol. *41*, 867 (1963)
178. Cox, C. S., Baldwin, F.: J. Gen. Microbiol. *49*, 15 (1967)
179. Van der Waals, J. H., Plateeuw, J. C.: Advanc. Chem. Physic *2*, 1 (1959)
180. Northcote, D. H.: Biochim. Biophys. Acta *11*, 471 (1953)
181. Akers, T. G.: Survival of virus, phage, and other minute microbes, in Introduction to Experimental Aerobiology, Dimmick, R. L. (ed.), p. 296. New York, N.Y.: Wiley-Interscience 1969
182. Rentschler, H. C., Nagy, R., Mouromself, G.: J. Bacteriol. *42*, 745 (1971)
183. Rentschler, H. C., Nagy, R.: J. Bacteriol. *44*, 85 (1942)
184. Gates, F. L.: J. Gen. Physiol. *13*, 231 (1929)
185. Zimmerman, L.: J. Bacteriol. *84*, 1297 (1962)
186. Fincher, E. L., Kethley, T. W., Cown, W. B.: Appl. Microbiol. *5*, 131 (1957)
187. Ringrose, R. E. et al.: Ann. Intern. Med. *69*, 719 (1968)
188. Sanders, C. V. et al.: Ann. Intern. Med. *73*, 15 (1970)
189. Traub, W. H., Raymond, E. A., Startsman, T. S.: Appl. Microbiol. *21*, 837 (1971)
190. Williams, R. P.: Appl. Microbiol. *25*, 396 (1973)
191. Tsang, J. C., Tattrie, S., Kallvy, D.: Appl. Microbiol. *22*, 1058 (1971)
192. Millikan, R. A.: Electrons (+ and −) Protons, Neutrons, Mesotrons, and Cosmic Rays, University of Chicago Press 1947
193. Marquis, R. E.: J. Bacteriol. *95*, 775 (1968)
194. Plummer, D. T., James, A. M.: Biochim. Biophys. Acta *53*, 453 (1961)
195. Riddick, T. M.: Basic but little recognized facets of zeta potential, Proc. Ind. Water Waste Conf., Dallas, p. 78, 1965
196. Schott, H., Young, C. Y.: J. Pharm. Sci. *61*, 182 (1972)
197. Haydon, D. A.: Biochim. Biophys. Acta *50*, 457 (1961)
198. Abramson, H. A.: Electrokinetic Phenomena and Their Application to Biology and Medicine, Chemical Catalog Co., New York, N.Y., 1934

Characterization and Performance of Single- and Multistage Tower Reactors with Outer Loop for Cell Mass Production

Karl Schügerl
Institut für Technische Chemie der Universität Hannover,
Callinstraße 3, D-3000 Hannover, FRG

There is a need to develop more and better methods to characterize cultivation media and bio-reactors.

The present paper reviews these methods using the example of cultivations of *Hansenula polymorpha* and *Escherichia coli* in bench-scale tower loop reactors under different operational conditions. The influence of medium properties, construction parameters, and process variables on the oxygen transfer rate and cell productivity are considered.

Though in the present state experimental data evaluated up to now are not sufficient enough to recommend general procedure for the layout, design, and scale-up of these bioreactors, some quantitative relationships are considered in this review also.

1 Introduction

Stirred tank reactors are the most important reactor types employed in biotechnology. However, since there is also a need for other reactors, several novel types have recently been developed for special manufacturing processes, which offer technical and economic advantages in comparison to stirred tank reactors. Especially large-scale single-cell protein production and waste water treatment require new reactor types. In addition, plant and animal tissue production needs suitable reactors. Several new reactor types were recommended in patents and in scientific literature [1]. However, reliable information on the properties and performance of these reactors is scarce. Tower and tower loop reactors belong to the few new reactor types which have already attained the level of production plant (50,000 t/a SCP of ICI in Billingham) or demonstration plant (1000 t/a SCP of Hoechst/Uhde in Hoechst) as well as that of large units for waste water treatment (20,000 m³ volume of Bayer in Leverkusen).

In spite of the considerable know-how which has been accumulated by these companies, very little key information has been published on these larger units.

Most of our knowledge is based on results evaluated using laboratory or bench-scale equipment. In Chapter 2, a short survey of these publications is given.

2 Survey of Tower and Tower Loop Reactors

The simplest tower reactor is a single-stage bubble column (1.1 in Table 1). To improve its performance different modifications are employed: stage separating trays (1.2), liquid pulsation (1.3), stage separating trays and rotating impellers (1.4), vibrating perforated plates (1.5) or motionless mixers (1.6). By these modifications the oxygen transfer rate, OTR, can be improved, but the power input is also increased. The aeration efficiency E_{O_2} (O_2 kg/kWh) of these modified reactors is usually lower than that of a single-stage tower reactor, if coalescence suppressing media are amployed [1]. To increase the residence time of the gas phase, countercurrent flow (1.7, 1.8) or concurrent downstream flow (1.9) of the phases is employed. When using a loop, the biomass concentration in the reactor is more uniform and the maximum gas load is much higher than in tower reactors without a loop. However, the volumetric mass transfer coefficient, $k_L a$, is considerably reduced. When using a coaxial draught tube (2.1, 2.2 in Table 2) or a vertical plane (2.3) or channel (2.4), the liquid degassing is incomplete. By means of an external loop (2.5, 2.6) and a horizontal connection of the loop and tower, degrassing can considerably by improved. Reduction of the loop's cross section reduces the liquid circulation rate (2.5 vs. 2.6).

To save compression energy during the operation of tall towers or deep shafts, the aeration is performed in the downcomer (3.1, 3.2). This is possible if, by aerating the upcomer section, the liquid circulation attains a velocity which is considerably higher than the bubble swarm velocity. This concurrent downstream flow can also be maintained after reduction of aeration in the upcomer section.

Several tower loop reactors are known which use a liquid pump to maintain the concurrent downstream flow. Such a reactor is the submerging channel (3.3), in which the liquid flow rate is accelerated and the static pressure is reduced in a nozzle at the top of the reactor. When employing slits at the narrowest cross section of the nozzle, self-aeration occurs. Especially if pure oxygen is being used, complete consumption of the gas is important. This is easily achieved in the downstream bubble column (3.4) and submerging jet reactor (3.5). In plunging jet reactors (3.6), a downcoming liquid jet is produced which hits the liquid pool surface, penetrates it, and carries air into the liquid.

Table 1. Tower reactors

	Concurrent upstream flow	Ref.
Single stage	1.1 —	2–42)
Multistage	1.2 —	43–56)
	1.3 Liquid pulsation	57–58)
	1.4 Rotating impeller	59–71)
	1.5 Vibrating perforated plates	70, 72–75)
Motionless mixer	1.6 —	55, 56, 76–79)
	Countercurrent flow	
Single stage	1.7 Liquid pump	87–90)
Multistage	1.8 Liquid pump	80–90)
	Concurrent downstream flow	
Single stage	1.9 Liquid pump	91, 92)

Table 2. Tower (shaft) loop reactors

		Concurrent upstream flow in the primary aerated section		Ref.
Internal loop	Coaxial draught tube	Air lift	2.1	112, 115 – 128, 130 – 138)
		Jet loop	2.2	93 – 114)
	Vertical plane	Air lift	2.3	129, 130, 133, 139 – 142)
	Vertical thin channels	Air lift	2.4	129)
External loop	Downcomer with large cross section	Air lift	2.5	143 – 151)
	Downcomer with small cross section	Air lift	2.6	14 – 27, 58, 153 – 155, 178)

		Concurrent downstream flow in the primary aerated section		
Internal loop	Coaxial draught tube	Air lift	3.1	130, 133, 139)
	Vertical plane	Air lift	3.2	130, 133, 139)
	Submerging channel	Liquid pump	3.3	156 – 159)
External loop	Downstream bubble column	Liquid pump	3.4	91, 92)
	Submerging jet	Liquid pump	3.5	160)
	Plunging jet	Liquid pump	3.6	161 – 177)

		Countercurrent flow in the aerated section		
External loop	Bubble column	Liquid pump	4.1	87 – 90)

Table 3. Tower loop reactors with trays or motionless mixers

Concurrent upstream flow in the primary aerated section				Ref.
Internal loop	Coaxial draught tube	Air lift	5.1	179)
	Vertical plane	Air lift	5.2	141)
External loop	Motionless mixer	Air lift	5.3	77)
	Trays	Air lift	5.4	17, 180)

Countercurrent flow in the aerated section				
External loop	Trays	Liquid pump	5.5	87 – 90)

The oxygen transfer efficiency of this reactor is, however, low [1]. Trays (5.1, 5.2, 5.4 in Table 3) or a motionless mixer (5.3), can improve the oxygen transfer rate of loop reactors. They also reduce the liquid circulation rate, but can improve the oxygen transfer efficiency if antifoam agents are employed. This is also true for countercurrent systems (4.1 in Table 2 vs. 5.5 in Table 3).

3 Aim of Review

In this review only single- and multistage tower loop reactors in concurrent and countercurrent operations with and without antifoam additives are treated as examples. Yeast (*Hansenula polymorpha*) and bacteria (*Escherichia coli*) cultivations in nonlimited substrate and oxygen transfer limited growth ranges as well as batch, fed batch and continuous reactor operations are considered. The influence of construction parameters and process variables on the oxygen transfer rate and cell productivity are considered. The importance of physical, chemical and biological, as well as reaction engineering characterization of the employed systems is stressed. Therefore, considerable information has been collected for a better understanding of the interaction of physical, chemical and biological processes. Unfortunately, the experimental data evaluated up to now are not sufficient enough to recommend quantitative relationships in the present status.

This paper intends to review the present state of our knowledge and encourages the development of more quantitative relationships for design and scale-up of tower loop reactors with low liquid circulation.

4 Materials and Methods

4.1 Characterization of Microbial Systems

Hansenula polymorpha (CBS 4732) and *Escherichia coli* (ATCC 11 105) were used for the reaction engineering measurements in the author's laboratory.

Hansenula polymorpha was investigated mainly with regard to the mechanisms of the oxidation of methanol by the enzyme system of this yeast [236-242]. Activities of methanol dissimilating enzymes and maximum specific growth rates were investigated on different carbon and energy sources only to compare them with the methanol system [240, 242]. These investigations show that μ_m is largest with glucose and those with ethanol and glycerol are nearly the same (Table 4).

Furthermore, one recognizes from Table 4 that glucose and ethanol repress the synthesis of methanol-catabolizing enzymes [240].

On the other hand, in glucose-limited chemostat culture increasing repression of all dissimilation methanol enzymes was found with decreasing dilution rate. It was also shown that not only high glucose concentration, but also nitrogen limitation causes repression of the synthesis of the dissimilatory methanol enzymes [242].

Very little data is available on *Hansenula polymorpha* using ethanol as a substrate [16, 87]. Table 5 shows a comparison of the maximum specific growth rates, yield

Table 4. Activities of the methanol-dissimulating enzymes and maximum specific growth rates of *Hansenula polymorpha* during growth on different carbon and energy sources [240]

Substrate	Max. specific growth rate μ_m (h^{-1})	Specific activities of			
		Alcohol oxidase	Catalase	Formaldehyde dehydrogenase	Formate dehydrogenase
Glucose	0.59	0.00	50	0.00	0.00
Sorbitol	0.53	0.31	440	0.18	0.02
Ethanol	0.33	0.00	300	0.00	0.00
Glycerol	0.31	1.92	900	0.52	0.16
Ribose	0.26	0.74	550	0.22	0.02
Methanol	0.13	3.10	1910	1.31	0.28
Xylose	0.07	1.74	720	0.48	0.01

Table 5. μ_m, $Y_{X,S}$, $Y_{X/O}$ and RQ of *Hansenula polymorpha* on glucose and ethanol substrates in limited growth phase in cocurrent bubble column [16]

Substrate	μ_m (h^{-1})	$Y_{X/S}$ (—)	$Y_{X/O}$ (—)	RQ (—)
Glucose	0.58	0.47	1.36	1.0
Ethanol	0.26	0.60	0.50	0.48

coefficients and respiratory quotients of *Hansenula polymorpha* with glucose and ethanol [16].

Escherichia coli

E. coli is one of the best investigated microorganisms. Its properties are described in several handbooks (e.g. [152]). Diderichsen found recently [314] that *E. coli* frequently changes its properties. Most of the *E. coli* stains can be considered heterogeneous. Two types of strains can be distinguished by their different sedimentation tendencies. One of them has a strong tendency toward flocculation and has the gene flu (fluffing, flocculating), the other one forms stable cell suspensions and has the gene flu$^+$ (nonfluffing, nonflocculating). A flu$^+$ colony always has 2–4% flu types and a flu colony, 8–12% flu$^+$ types. The frequent change of their ratio indicates the metastability of the gene. This gene controls cell flocculation by variation of the surface properties of *E. coli*. This property is important for cultivation of *E. coli* in tower reactors as well as for their separation from the medium without employing a centrifugal separator.

Determination of the Biological Parameter

Contrary to the behavior in well-mixed continuous stirred tank reactors, in which the spacial variations of pressure, cell-, substrate- and oxygen concentrations are moderate, the cells in tower loop reactors are exposed to considerable pressure and concentration variations. If the longitudinal liquid mixing intensity is low, the oxygen and substrate concentrations can considerably vary along the tower as well as in the loop. Therefore, the influence of periodical variation of the cell environment on cell behaviour has to be taken into account. The use of biological parameters evaluated in small continuous laboratory reactors under steady state conditions in a uniform cell environment is generally not transferable to larger units. Since the influence of the actual conditions on cell behaviour cannot be predicted, it is necessary to determine the biological parameters under operational conditions.

Only as long as uniform conditions in the reactor prevail, is it possible to use the well known mass balances of cell, substrate and oxygen, which are valid for reactors with lumped parameters. In reaction with distributed parameters, no simple relationship holds true, although some simplifications are possible.

a) According to the observations in the author's laboratory uniform cell concentration prevails in bench scale tower loop reactors, as long as the liquid circulation rate exceeds a critical value. Thus the maximum specific growth rate, μ_m, can be calculated under these conditions in the nonlimited growth range from the variation of (dry) biomass, X, with time in batch fed batch reactors:

$$R_X^* = \frac{dX}{dt} = \mu_m X,$$

(1)

or in continuous reactors under steady state conditions and in the nonlimited growth range (which is not a stable state in contrast to the substrate-limited state)

$$\mu_m X = DX .\tag{2}$$

In substrate or oxygen transfer limited growth, the specific growth rate, μ_m, can usually be calculated by the twofold Monod model:

$$\mu = \mu_m \frac{S}{K_s + S} \frac{O}{K_o + O} .\tag{3}$$

Since in tower loop reactors in general, $S(z, t)$ and $O(z, t)$ are valid, i.e. substrate concentration, S, and dissolved oxygen concentration, O, depend on the length coordinate, z, and on the time, t, the specific growth rate μ is also space and time-dependent: $\mu(z, t)$. It is difficult to determine the (saturation) constants: K_s (with regard to the substrate) and K_o (with regard to oxygen). However, recent results indicate that the assumption of time and space independent K_s and K_o constants is fairly realistic.

b) The process simulation results [181] indicate that under oxygen transfer limiting conditions, the substrate concentration, S, is nearly constant in bench scale tower loop reactors. Thus the tower loop reactor behaves not only with regard to the cell but also with regard to the substrate concentration as a lumped parameter reactor. Thus in extended culture (fed batch culture with constant substrate concentration) only the spacial dependence of the dissolved oxygen concentration, $O(z, t)$, must be taken into account.

The (dry) biomass concentration, X, was measured gravimetrically as a function of time. Thus μ_m was determined by Eq. (1) in the nonlimited growth range in batch or fed batch cultivations. The determination of μ_m in a continuous culture in the nonlimited growth range, under steady state conditions by means of Eq. (2) is only possible in a reactor with an on-line measurement of the cell density and with control of the dilution rate, D, to keep X constant. Since no instrument is available which can measure X on-line, Eq. (2) can only be used to determine μ_m if it is possible to calculate X on-line by means of a mass balance of the components, e.g., in the simplest case from the CO_2 concentration in the outlet gas.

In lumped parameter reactors, the substrate and oxygen yield coefficients, $Y_{X/S}$ and $Y_{X/O}$, can be calculated from the uniform cell growth, substrate, and oxygen consumption rates by means of Eqs. (4) and (5):

$$Y_{X/S} = \frac{R^*}{R_S}\tag{4}$$

$$Y_{X/O} = \frac{R_X^*}{R_O}\tag{5}$$

where $R_S^* = -\dfrac{dS}{dt}$, the substrate uptake rate and

$R_O^* = -\dfrac{dO}{dt}$, the oxygen uptake rate.

However, since, in general, R_X^*, R_S and R_O are space and time dependent, $Y_{X/S}$ and $Y_{X/O}$ can only be calculated by means of cell, substrate and oxygen mass balances in the reactor and by fitting the calculated spacial and time variations of $X(z, t)$, $S(z, t)$ and $O(z, t)$ to the measured ones.

a) In bench scale tower loop reactors X is space independent and because of the slow variation of X in comparison to the variations of S and O a quasi-steady state can be assumed. Thus for a short time, $R_X^* = $ const. can be assumed.

b) Under oxygen transfer limiting conditions in extended culture, X = const. can be assumed and Eq. (4) can be used.

To determine $Y_{X/O}$, $O(z, t)$ has to be calculated and measured.

In the tower loop reactors we employed the dissolved oxygen partial pressures along the tower in 13 different positions on-line, thus $O(z, t)$ was available. Furthermore, highly sophisticated distributed parameter models were used to calculate $O(z, t)$. These allowed the determination of $Y_{X/O}$ by fitting the calculated dissolved oxygen concentration profiles to the measured ones, by means of nonlinear optimization methods (see 5. Models).

Because of the spacial and time dependences of the environment of cells in tower loop reactors, it would be necessary to employ structured models. When using unstructured models (e.g. Monod models), the biological parameters can exhibit dependence on reactor types or operational modes (see 7.3).

4.2 Methods for the Characterization of Media

4.2.1 Salt Effects

It is well known that salts considerably influence medium properties, especially cell flocculation [249], foam formation [183] and bubble coalescence [13, 185−188]. However, these properties are influenced not only by the ionic strength I*

$$I^* = \frac{1}{2} \Sigma \, c_i z_i^2 \, , \tag{6}$$

where c_i is the ionic concentration and
z_i the ion valency,
but also by the particular ions themselves.

According to Bernal and Fowler [189] ions affect the neighbouring water molecules by loosening or tightening the structure. A looser structure due to unhydrated ions decreases the viscosity below that of pure water. On the other hand, a tighter structure gives rise to higher viscosity. Relatively small and multivalent ions increase the viscosity of water by polarizing water molecules even beyond the first water layer (structure makers) [190]. Also anions such as Cl^- and Br^- have small structure breaking properties and actually make the water more fluid. However, the presence of highly charged anions e.g. SO_4^{2-}, in $MgSO_4$ is responsible for its greater effect (increase) on viscosity as compared with $MgCl_2$, which makes the former a more powerful structure maker. The structure making or breaking properties of ions are given in the so called Hofmeister lyotropic series [191−193]. The position of a given salt in this series can also be determined experimentally by means of the turbidity temperature [183]. In Table 6 a series of salts are listed according to their colloid "salting out" effect. The salts at the top have the strongest structure breaking (salting in) effect and those at the bottom the strongest structure making (salting out) effect. The position of the employed nutrient solution on this series was determined by the turbidity temperature method.

The spatial composition variation of the nutrient salt solution in the reactor is negligible.

Table 6a. Turbidity temperatures, T_T of aqueous solutions of C_8H_{17}⟨◯⟩$-(O-CH_2-CH_2)_9OH$ (PIOP-9) (10 g l^{-1}) with 0.5 m salt addition. T_T without ions: 64 °C [193]

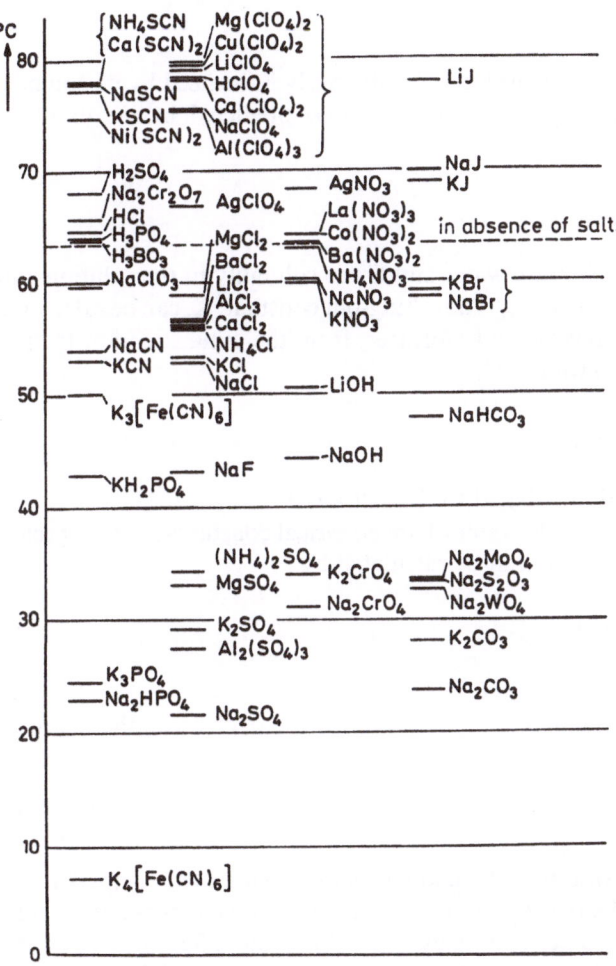

4.2.2 Substrate Effects

Ethanol and glucose were employed as substrate. Alcohol possesses a surface tension much lower than that of water. The surface tension of an alcohol-water solution lies, of course, between the surface tension of pure water and alcohol and depends on concentration. Alcohol concentrations influence bubble coalescence [13, 194, 195], foaming [196], as well as mass transfer coefficients, k_L [197]. Alcohol concentrations in the medium were determined by gas chromatography.

Glucose increases the surface tension of water because, similar to inorganic salts, it makes the solution more cohesive and hence it promotes the formation of bound (highly structured) water in sufficiently high concentration. Under these conditions excess water molecules, and not the solute, adsorb on the surface. Glucose does not influence the bubble coalescence appreciably.

At high glucose concentrations, the employed yeast produces ethanol. The glucose concentration was measured by a polarimeter (PM 241, Perkin-Elmer).

4.2.3 Oxygen and CO_2 Solubilities

Oxygen and CO_2 solubilities in the medium are obviously influenced by its composition. In simple salt solutions these solubilities can be predicted by means of the Sechenov equation.

$$\log (C_0/C) = KC_{el} \tag{7}$$

where C_0 is the gas solubility in pure water and C the solubility in the solution with a molar concentration of electrolyte C_{el}. The salting-out constant, K, can be calculated, according to the model of Krevelen and Hoftijzer, from the ionic strength, I_i^*. This model was modified by Danckwerts [198]:

$$\log (C_0/C) = \Sigma\, h_i I_i^* \tag{8}$$

where I_i^* is the ionic strength attributed to the salt i and
$\qquad h_i = h_G$ (gas); h_+ (cation), h_- (anion) are empirical constants for the species i
According to Deckwer [199, 200], C can be calculated by

$$\log (C_0/C) = \Sigma\, H_i I_i^* \tag{9}$$

The constants H_i are tabulated in Ref. [200].
When considering organic compounds (glucose, saccharose, glycerol) the following model is recommended [20]:

$$\log (C_0/C) = a + b\, (\Sigma\, H_i I_i^*) \tag{10}$$

The constants are tubulated in Ref. [201] and oxygen solubilities for complex media are given in Ref. [202]. Since the oxygen solubility in the medium changes during cultivation [203], it is necessary to determine it during cultivation. These measurements were carried out by the research group of Deckwer, employing a closed thermostatized system, in which, after medium degassing, the oxygen and/or CO_2 pressure was measured above the medium until a constant value was attained [199, 200].

4.2.4 Diffusivity of Dissolved Oxygen and Protein

Diffusion coefficients are determined by means of the gas absorption rate under defined fluiddynamical conditions (laminar jet, laminar film on cylinder, laminar film on sphere etc.).
When employing the model of Heyduk and Chang [204], the diffusivity can be calculated by

$$D_M \eta^A = K_1 \tag{11}$$

where D_M is gas diffusivity in the medium,
   η the dynamical viscosity of the medium and
   A and K_1 empirical constants.
For $\eta < 1.5$ mPas $A \simeq 1$ and for
   $\eta > 5.0$ mPas $A \simeq 0.66$ [205)]
   Diffusivity of protein influences its enrichment on the interface as well as foam and
small bubble coalescence properties. It can be calculated by Eq. (12) [206)]:

$$D_M = 8.34 \times 10^{-8} \, (T/\eta^* M^{1/3}) \tag{12}$$

where M is the molecular weight of the protein
   T the absolute temperature and
   η^* the solvent viscosity.

4.2.5 Viscosity of the Medium

Cultivation media of yeast and bacteria exhibit Newtonian behaviour. Therefore it
is not necessary to measure the entire shear diagram for the determination of the
dynamic viscosity, η. For yeast (*Saccharomyces cerevisiae* and *Candida utilis*) suspensions η can be calculated by Eq. (13) [207)]:

$$\frac{\eta_s}{\eta_0} = \frac{1}{1 - (h_s \varepsilon_x)^a} \tag{13}$$

where η_s is the viscosity of the suspension
   η_0 the viscosity of the supernatant
   ε_x the volume fraction of the cells
   h_s the packing factor:
      $h_s = 0.0487 \, p_{osm} + 1.59$ and
   p_{osm} the osmotic pressure (bar $= 10^5$ N m^{-2})
   a   empirical constant
   The viscosities of cultivation media were measured by a rotational viscosimeter
(Rotovisco RVII, Haake, Berlin).

4.2.6 Interfacial Properties

It is well known that interfacial properties can influence mass transfer across the
interface [208)].
   The most common characterization of the liquid surface is by its surface tension,
σ. In biological media, however, σ is time dependent and it takes about 10 to 20 h to
attain the equilibrium surface tension [183, 196, 209, 210)]. Therefore, most of the published
surface tension data on biological media is useless, since surface age information is
missing.
   Additive information can be gained by a frequency response method [211, 212)].
   The amplitude decrease and phase shift of capillary and/or longitudinal waves,
which are produced on the liquid surface, are measured as a function of the distance
from the wave generator. The surface viscosity and/or elasticity were evaluated from
these measurements for protein solutions [213, 214)].

These measurements indicate that the presence of protein in the medium increases the surface viscosity considerably. The surface behaves like a viscoelastic body at high protein concentrations.

4.2.7 Bubble Coalescence Promoting or Hindering Effect of the Medium

The interfacial properties considerably influence the bubble coalescence process in liquids, which is a very complex process.

In aqueous solutions it depends on the water structure [184–187, 215]. In alcohol solutions it depends on $d\sigma/dc$, where σ is the surface tension (mN m^{-1}) and c is the solute concentration (mol kg^{-1}), and on ΔP, which is the excess internal pressure in the film [195]:

$$\Delta P = \frac{2\sigma}{r} + \frac{A_0}{6\pi h^3} - 64 R_G T C_{el} \gamma^2 \exp(-\varkappa h)$$ (14)

where A_0 is the Hamaker-London constant for water (3.8×10^{-20} J)
  R_G the gas constant
  r   radius of bubble
  F   the Faraday constant
  C_{el} the electrolyte concentration
  K   the Debye reciprocal length
  h   the thickness of the film lamelle
  γ the tanh $(zF\Phi_0/4R_G T)$
  z   the ion valency and
  Φ_0 the double layer potential at the phase boundary
If the film is assumed to stretch elastically, i.e. the volume in the film is conserved, then the specific surface expansion rate, S_s, is given by

$$S_s = \frac{1}{A}\frac{dA}{dt} = -\frac{1}{h}\frac{dh}{dt}$$

The time, t'_s, for stretching a film from infinite thickness to a final thickness, h_f, is then given by Eq. (15):

$$t'_s = \frac{\pi}{2DR_G^2 T^2} \int_{\infty}^{h_f} \frac{[2c(d\sigma/dc)^2 + h\,\Delta P(d\sigma/dc)]^2}{h^3(\Delta P)^2}\,dh$$ (15)

When substituting Eq. (14) into (15), the complete expression for t'_s is obtained. The numerical integration of Eq. (15) gave t'_s as function of h_f. These expressions agree fairly well with the measurements carried out employing n-alcohols [195]. Because of this complex interrelationship between liquid properties and coalescence parameters, coalescence effects were often measured indirectly by means of determining E_G, d_s, a or $k_L a$ through variation of the liquid in the same equipment and under the same operational conditions [13, 184–188, 215–218].

At the same concentration and chain length, the fatty acids have the strongest, the alcohols intermediate, the poly-alcohols and ketones the slightest influence. The concentration, C_{co}, at which the coalescence suppression begins, is inversly proportional to the number of carbon atoms, n_c, [216]:

$$C_{co} \propto n_c^{-1.5}$$ (16)

The interfacial tension suppression, $d\sigma/dc$, is also influenced by C_{co} [216]:

$$\ln \frac{d\sigma}{dc} = -(1.5 + 0.5 \ln C_{co}) \tag{17}$$

There is a definitive relationship between the coalescence suppression effect of salt solutions, their ion strength and their position in the lytropic (Hofmeister) series. At the same ion strength, the salt has a stronger effect which exhibits a greater tendency to flocculate proteins (see 4.2.1).

Since no relationship is known for complex cultivation media the bubble coalescence behaviors of the media were experimentally determined by means of the volumetric mass transfer coefficient, $k_L a$, which was measured in a standard bubble column reactor. These $k_L a$ values were compared with the $k_L a$ values measured in the same equipment under the same operating conditions with water: $(k_L a)_{H_2O}$. The ratio, m:

$$m = \frac{k_L a}{(k_L a)_{H_2O}} \tag{18}$$

was employed to characterize the cultivation medium for a fixed superficial gas velocity, w_{SG} [217]. In the presence of antifoam agents, Eq. (18) was modified:

$$m^* = \frac{(k_L a)_{corr}}{(k_L a)_{ref}} \tag{19}$$

As a reference, the nutrient salt solution with antifoam agent (Desmophen 3600; Bayer AG) was employed. The $k_L a$ values measured in the investigated media were corrected by considering the differences in k_L values in the investigated and reference media at low superficial gas velocities, at which no coalescence is present [218]:

$$(k_L a)_{corr} = k_L a - \Delta(k_L a) \tag{19a}$$

$$\Delta(k_L a) = \left[k_L a - (k_L a)_{ref}\right]_{w_{SG} = 2\ cm\ s^{-1}} \tag{19b}$$

4.2.8 Foaminess and Foam Stability

Cultivation is often accompanied by foam formation due to the high foaming capacity of protein solutions. This capacity results from the stabilization of the gas liquid interface caused by the denaturation and strong adsorption of the surface proteins [219].

Since mechanical foam breaking with its high input power requirement is expensive, antifoam agents are usually preferred in the bioindustry. The presence of antifoam agents, however, decreases the efficiency of gas dispersion by increasing the bubble coalescence rate. To compensate, the specific power input rate must be increased, which again increases production costs [220]. Therefore, foam formation can considerably influence production costs.

The influence of salts [183] and alcohols [196] on foaminess were investigated and simple relationships were found. The salt effect can be explained by the interaction between the water and salt, i.e., by its influence on the water structure (see 4.2.1). The addition of salts to the protein solutions enlarged

Table 6b. Constant of Eq. 20

Salt	k °C l mol^{-1}	T_{T0} °C	T_{corr} °C
KCl	− 18.0	60.8	0.0
MgSO$_4$	− 68.0	60.5	0.3
K$_4$Fe(CN)$_6$	−125.0	60.8	0.6^a
			2.4^b

^{a} for 5.0×10^{-2} g l^{-1} BSA solution
^{b} for 1.0×10^{-1} g l^{-1} BSA solution

their foaminess, Σ, due to the increasing apparent protein concentration caused by the change in water structure.

The change of foaminess can be calculated as follows [183]: Turbidity temperature of the salt solution, T_T, can be measured or calculated by Eq. (20):

$$T_T = T_{T0} + C_{salt}k \qquad (20)$$

where T_{T0} is the turbidity temperature of the salt free NP-10 solution extrapolated to $C_{salt} = 0$ and
    k   a specific constant for the applied salt, calculated from the slope of $T_T(C_{salt})$ (Table 6b)
From T_T and T_{T0}, their difference:

$$\Delta T_T = T_{T0} - T_T$$

can be calculated. In some cases better results can be obtained when T_{T0} was corrected by addition of T_{corr}, which takes into account the difference between T_{T0} in the salt free NP-10 solution (10 g l^{-1}) and the extrapolated T_T value at $C_{salt} = 0$. Inserting ΔT into Eq. (21):

$$C_1 = C_0 2^{\Delta T_T}, \qquad (21)$$

where C_0 is the actual concentration of the protein in the solution and
    C_1 its apparent concentration,
   C_1 was determined. From the diagram $\Sigma = f(C)$, and the new foaminess is given in C_1 instead of C_0.

Alcohols cause an increase of the turbidity temperature of tensides. Additives, which increase the turbidity temperature of detergents should diminish the effective concentration and protein foaminess only if the variation in water structure plays an important role, but alcohol additives increase protein foaminess [196]. This means that the direct interaction between alcohols and proteins overcompensates the influence of water structure variation on the foaminess due to the alcohol additives.

These two effects can be taken into account by means of a simple relationship [196].

At low alcohol concentrations, a linear relationship between turbidity temperature (with NP-10 solution), T_T, and alcohol concentration prevails:

$$T_T = T_{T0} + c_m k \qquad (22)$$

or

$$\Delta T_T = -c_m k \qquad (23)$$

where c_m is the molar alcohol concentration,
    k   a constant and
   T_{T0} the turbidity temperature at $c_m = 0$

Table 6c. Constants a and n of Eq. (25)

Alcohol	BSA (g l^{-1})	a	n
Methanol	5.0×10^{-2}	3.92	0.55
Ethanol	5.0×10^{-2}	5.09	0.48
Propanol	5.0×10^{-2}	5.00	0.26
Methanol	1.0×10^{-1}	4.29	0.51
Ethanol	1.0×10^{-1}	5.56	0.36
Propanol	1.0×10^{-1}	5.35	0.25

However, since alcohol additives increase the effective concentration of protein and do not, as expected, decrease it, the apparent concentration variation was taken into account by an additive term, which considers the direct alcohol-protein interaction:

$$\Delta T_{T\,eff} = \Delta T_T + \Delta T_D \tag{24}$$

where $\Delta T_{T\,eff}$ is the effective temperature difference,
$\quad \Delta T_T$ the turbidity temperature difference due to pure solvent structure effect and
$\quad \Delta T_D$ a correction term for direct alcohol protein interaction.
The concentration dependence of ΔT_D can be described by Eq. (25):

$$\Delta T_D = a(c)^n \tag{25}$$

where c is the alcohol concentration (% v/v) and a and n are constants (Table 6c).
The protein concentration correction factor, f, is given by Eq. (26):

$$f = 2^{\Delta T_T T_{eff}} \tag{26}$$

If the foaminess, Σ, of the pure protein solution is known as a function of the protein concentration, the foaminess of the protein solution in the presence of an alcohol can be calculated by correcting the protein concentration c_{prot} by f.
The new foaminess value, Σ, can be taken from the $\Sigma(c_{prot})$ diagram at the new concentration $(c_{prot}f)$.

However, it is usually not possible to calculate the foaminess of complex cultivation media by these simple relationships. Therefore Σ must be experimentally determined [220]. The measurements of foaminess, Σ, were carried out at 20 °C according to Bikermann [221] in glass tubes 5.5 cm in diameter and employing a porous plate (G4-frit) 2.5 cm in diameter for values of $\Sigma > 50$ s and in glass tubes 2 cm in diameter and a porous plate 1.2 cm in diameter for $\Sigma < 50$ s, respectively. The distance between the lower foam boundary and the porous glass surface was kept constant (1 cm). Foaminess, Σ, was defined by Eq. (27):

$$\Sigma = \frac{V_s}{V_{tg}} \tag{27}$$

where V_s is the equilibrium volume (cm^3) of the foam above the liquid layer,
$\quad V_{tg}$ the volumetric gas flow rate (cm^3 s^{-1}).
By employing this method, foaminess of yeasts [210] and fungi [209] cultivations were determined.

Foam stability can play an important role if the foam is to be eliminated by mechanical foam destroyers. A simple method to determine foam stability was worked out by Bumbullis [213, 222]. A glass paddle stirrer (52 mm in width, 20 mm in height and 2 mm in thickness) was installed above the foam column. With a rotational speed of 0.45 rpm the column was slowly raised during which the foam at the surface of the column was destroyed. As soon as the foam was no longer destroyed by the paddle, but began sticking to it, the column elevation was stopped and the remaining foam volume, V_{sR}, measured and the remaining foamines, Σ_R, calculated by Eq. (28):

$$\Sigma_R = \frac{V_{sR}}{V_{tg}} \tag{28}$$

Σ_R is the remainder of the foaminess, which was not destroyed by the mechanical action of the paddle, i.e. was stable under these conditions.

The ratio of the remainder and original foaminess

$$S^* = \frac{\Sigma_R}{\Sigma} \tag{29}$$

was employed to characterize the relative foam stability, S^*.

Whereas foaminess is increased in the presence of salts, the same salts cause a diminution of relative foam stability S^*. For each salt, a factor f_b can be defined, by which the apparent protein concentration in the solution is altered with regard to the foam stability. For the average value of the products of f (protein concentration correction factor due to salt effects with regard to the foaminess of the protein solution) and f_b (protein concentration correction factor due to salt effects with regard to the relative foam stability, S^*) Eq. (30) holds true

$$ff_b = 1.19 \tag{30}$$

More detailed investigations indicate that ff_b is a slight function of the protein concentration [222]. In general, one can state that as the foaminess of protein solutions in the presence of salts increases their relative stability diminishes. These two effects nearly compensate each other and can be described quantitatively by means of the turbidity temperature shift due to the interaction between the salts and the water structure.

The influence of alcohol additives on the foam stability could not yet be determined because of the relatively short life times of foams in the presence of alcohols.

4.2.9 Protein and Cell Flotations

In the case of foam formation, cells can be enriched in foam. This is caused by foam flotation or micro-flotation of cells and can significantly reduce cell productivity in bioreactors due to the reduction of cell concentration in the medium. Cells in foam are usually in substrate limited state. Thus their specific growth rate is very low, so that foam flotation or microflotation of cells in bioreactors is not desired.

The yeast and bacteria cells, which were used for our investigations, cannot be foam flotated without additives because their surfaces are hydrophilic. By means of different additives their flotation (microflotation) is possible [243–247]. The proteins, which are present in the cultivation medium acts as additives. Thus *Hansenula polymorpha* can be microflotated in its original cultivation medium depending on the cultivation mode.

Cells in nonlimited growth state cannot be foam flotated without additives. Cells in substrate limited, especially in oxygen transfer limited states, can be flotated, if no antifoam agent is added to the system.

The cell flotability is determined experimentally by means of laboratory scale flotation equipment, which consists of a 500 mm high glass column, 21.6 mm in diameter with a thermostatized mantle. The foam is formed on the top of the bubbling layer consisting of 100 ml liquid and aerated with 1 ml s^{-1} of water saturated and thermostatized nitrogen. The foam is discharged through a bend and destroyed by means of a small mechanical foam destroyer [247]. The enrichment of the cells are characterized by the cell concentration ratio, X_s/X_p, and the cell enrichment ratio X_s/X_R, where X_s is the cell concentration in the foam liquid in the sample, X_p, and X_R in the rest (foam free liquid).

A significant relationship was found between the cell enrichment and protein enrichment ratios. Therefore protein enrichment was also determined in the same equipment and characterized by the protein concentration ratio P_S/P_P and enrichment ratio P_S/P_R, where P_S is the protein concentration in the foam liquid in the sample, P_p, and P_R in the rest (foam free liquid).

4.2.10 Cell Flocculation and Sedimentation

There is significant disagreement among different research groups with regard to the cause of cell flocculation [248-255]. The flocculation of yeast seems to be genetically determined. However, there is no flocculation during the non-limited exponential growth phase. Ions, especially Ca ions, phosphates, properties of cell membranes as well as glycogen content of the cells can influence the flocculation of yeasts during growth. Additives, which influence the water structure and protein solubility can alter the flocculation properties.

Since the influence of different additives on cell sedimentation based on literature data are contradictionary, it was necessary to determine experimentally the cell sedimentation rates. For cells with low sedimentation rates, batch runs were carried out. Two 70 cm high plexiglass cylinders, 14 cm in diameter, were used for these measurements. They had 9 sampling ports (at 7 cm distances) along the cylinder. The sampling ports consisted of rubber membranes attached to the cylinder wall. Through them probes could be taken by means of hypodermic syringes. The cell mass and protein concentrations were determined in these probes as functions of time. Furthermore, the displacements of the interfaces between layers of different cell concentrations along the column were determined optically. Because the displacement of the clear liquid/turbid liquid interface was largest, the sedimentation rate was calculated from the displacement of this interface. To compensate for the volume change due to sampling, the longitudinal distance was always related to the corresponding overall height of the layer. Another method was also employed to characterize the sedimentation rate. The cell concentration (extinction) variation in the probes with time, taken halfway up the cylinder, was employed to calculate the time necessary to reduce the cell concentration (extinction) to one half of its original value, $t_{1/2}$.

For flocculating cells with high sedimentation rates, continuous runs were carried out. The cell suspension was continuously fed into the sedimentation tank and, from the top, the clear liquid and, from the bottom, the concentrated cell suspension were continuously removed. The cell mass enrichment was measured as a function of the mean cell residence time in this sedimentation tank [256].

4.3 Methods for Characterization of Two-phase Systems

Two phase system properties can strongly influence cultivation conditions, especially if growth is oxygen transfer limited. To treat oxygen transfer rate quantitatively, it is necessary to determine the following properties: E_G, d_S, a, $k_L a$, u_B, $\bar{u}$. The primary variable here is the Sauter bubble diameter, d_s, because it influences the other variables. In all industrial bioreactors, gas dispersion is caused by turbulence [1]. The gas dispersion is closely related to the turbulent energy dissipation mechanism.

Turbulent flow produces primary eddies which have a wave length (or scale) of similar magnitude to the dimensions of the main flow. These large primary eddies are unstable and disintegrate into small eddies until all their energy is dissipated by viscous flow. Most of the kinetic energy is contained in the large eddies, but nearly all of the dissipation occurs in the smallest eddies. The properties of these small eddies are determined by the local energy dissipation rate per unit fluid mass [300]. If the scale of these eddies is much smaller than the bubble diameter, the bubbles will be dispersed. The maximum stable bubble size depends on specific power input, P_L/V_L.

Because of the eminent role of the turbulence energy dissipation in gas dispersion, its study is important. Therefore, the turbulence properties of two phase systems will be considered also in this review.

4.3.1 Relative Gas Hold up

The mean relative gas hold-up, E_G, is defined by Eq. (31)

$$E_G = \frac{V - V_L}{V} = \frac{H - H_L}{H}$$

(31)

where V   is the volume of the bubbling layer
      V_L the volume of the bubble-free liquid layer
      H   the height of the bubbling layer
      H_L the height of the bubble-free liquid layer.

The determination of E_G is usually carried out by measuring H and H_L in the tower. In tower reactors with outer loops, H is given by the reactor construction. H_L is calculated from the hydrostatic pressure, which is measured by a gauge at the bottom of the reactor just above the aerator.

The local relative gas hold up was measured by electrical conductivity probes (see 4.3.2).

4.3.2 Bubble Sizes

In tower loop reactors, the bubble size usually covers a wide range. If the bubbles are small, they have spherical shapes, and the bubble diameter, d_B, is used to characterize the bubble size. Because of the bubble size distribution, mean $\bar{d}_B$ or Sauter mean d_s, are calculated by:

$$\bar{d}_B = \frac{\sum\limits_1^N n_i d_{Bi}}{\sum\limits_1^N n_i} \tag{32}$$

$$d_s = \frac{\sum\limits_1^N n_i d_{Bi}^3}{\sum\limits_1^N n_i d_{Bi}^2} \tag{33}$$

where n is the frequency of the bubbles with the diameter, d_{Bi}.

The prerequisite for the determination of $\bar{d}_B$ and d_s is the knowledge of the bubble diameter distribution.

Several methods have been recommended in the last 20 years to determine bubble size distributions [223-229, 231]. The most advanced method of Burgess and Calderbank [223] cannot be applied in common tower reactors with biological media, since it cannot detect small bubbles which are present in these systems. By miniaturizing electrical conductivity probes bubble diameters larger than 0.6 cm can be realiably determined [226, 227]. Optical probes (dip-sticks), which use fibre optics to detect bubbles [231], electro-optical methods [226-229] as well as photographic methods [226] were also employed and compared [226, 227]. The reported bubble size distributions were measured by miniaturized electrical conductivity probes [226, 227].

The measurements with these probes also yield the local gas hold up in tower reactors [226, 227].

4.3.3 Bubble Velocities

Integral bubble velocities (swarm velocity), w_{BS}, can be calculated according to Eq. (34):

$$w_{BS} = \frac{w_{SG}}{E_G} \tag{34}$$

or by relative swarm velocities, w_{BS}, with regard to the mean liquid velocity by Eq. (34a)

$$w_{BR} = \frac{w_{SG}}{E_G} - \frac{w_{SL}}{1 - E_G} \tag{34a}$$

where w_{SG} is the superficial gas velocity and
w_{SL} the superficial liquid velocity.

Local bubble velocities were measured by miniaturized electrical conductivity probes [226, 227].

4.3.4 Specific Interfacial Areas

The specific interfacial area with regard to the liquid volume, a, can be measured directly by a light transmission technique [230, 232], by chemical methods [233−235] or calculated by Eq. (35),

$$a = \frac{6E_G}{d_s(1 - E_G)} \tag{35}$$

if the bubbles are small enough to have a nearly spherical shape.

Since light transmission and chemical methods cannot be employed in cultivation media, a was calculated by Eq. (35). Because of the small bubbles present in cultivation media, Eq. (35) is a good approximation.

To determine the local specific interfacial area, ε_G and d_s were measured by double electrical conductivity microprobes (see 4.3.2).

4.3.5 Volumetric Mass Transfer Coefficients

In reactors with a uniform concentration of substrate, cells and dissolved oxygen (reactors with lumped parameters), $k_L a$ can be calculated in batch operation by Eq. (36):

$$k_L a = \frac{OTR}{O^* - O} \tag{36}$$

and in continuous operation by Eq. (37):

$$k_L a = \frac{OUR - D(O_0 - O)}{O^* - O} = \frac{OTR}{O^* - O} \tag{37}$$

The oxygen transfer rate, OTR, was calculated from the O_2 balance of the gas phase by means of the gas compositions (O_2, CO_2, N_2) at the gas reactor inlet and outlet.

The oxygen utilisation rate, OUR, was calculated by Eq. (38):

$$OUR = \frac{1}{Y_{x/o}} \mu(X - X_0) \tag{38}$$

where X_0 is the cell concentration in the feed.

The saturation concentration of dissolved oxygen, O^*, was calculated by means of measured gas composition in the reactor and measured O_2 solubility in the medium. The dissolved oxygen concentrations in the feed, O_0 and reactor, O, were calculated by means of the measured dissolved oxygen tensions, p_0 and p, and the measured oxygen solubility in the medium. The dilution rate, D, was also measured. In reactors with nonuniform concentrations of substrate, cells and dissolved oxygen (reactors with distributed parameters), the determination of $k_L a$ is difficult, since in general all of the reactor parameters and process variables of Eqs. (36) and (37) are space dependent.

For the investigated tower loop reactors it was possible to assume that $X = \text{const.}$ The assumption $S = \text{const.}$ only holds true if oxygen transfer limited growth prevails.

In general $S(z)$, $O^*(z)$, $O(z)$ and $k_L a(z)$ had to be taken into account. $O(z)$ was measured, $S(z)$ was calculated by means of the substrate balance in the liquid and $O^*(z)$ by means of O_2, CO_2 and N_2 balances in the gas and liquid phases [181].

The assumption of CSTR or PFR with regard to the longitudinal dissolved gas concentration profile causes large error in the determination of $k_L a$ [76].

4.3.6 True Mean Liquid Velocities

The liquid velocity, u, is usually represented by Eq. (39):

$$u = \bar{U} + u_x + u_y + u_z \tag{39}$$

where $\bar{U}$ is the true mean liquid velocity and
u_x, u_y, u_z are the fluctuation components in the space directions x, y and z.

When employing wedge shaped probes [257], which are parallel with $\bar{U}$ the true liquid velocity can be evaluated from the mean values of the measured signals in the upstream direction, u_{up}, and in the downstream direction, u_{down}, by their addition [258, 259]:

$$u_{up} = \bar{U} + u_x + |u_y| \tag{40}$$

$$u_{down} = u_x + |u_y| \tag{41}$$

hence

$$\bar{U} = u_{up} - u_{down} \tag{42}$$

Equation (42) holds true, since wedge shaped probes record only positive values of u_x and u_y and do not record u_z (parallel to the edge).

4.3.7 Turbulence Intensity

When wedge shaped probes are used, the signal variance yields the following turbulence intensity

$$u'_{x, y, i} = \sqrt{\overline{(u_x + |u_y|)^2}} \tag{43}$$

When measuring $u'^2_{x, y\ up}$ and $u'^2_{x, y\ down}$ parallel with $\bar{U}$, a mixed turbulence intensity can be evaluated [259, 260]

$$u'_{x, y} = \sqrt{u'^2_{x, y\ up} + u'^2_{x, y\ down}} \tag{44}$$

Lippert has shown [261] that it is possible to evaluate u'_x by means of a cylindrical shield around the probe:

$$u'_x = \sqrt{\overline{u^2}} \tag{45}$$

The relative turbulence intensities are defined by Eqs. (45) and (46):

$$I_{x,y} = \frac{u'_{x,y}}{\bar{U}} \qquad\qquad (46\,a)$$

$$I_x = \frac{u'_x}{\bar{U}} \qquad\qquad (46\,b)$$

4.3.8 Autocorrelation Function of Velocity Fluctuations

Cross correlation function of velocity fluctuations, $R_{A,B}$ is suitable to characterize the turbulence structure. It is defined by Eq. (47):

$$R_{A,B} = \frac{\overline{u_{x,A} u_{x,B}}}{u'_{x,A} u'_{x,B}} \qquad\qquad (47)$$

Here $u_{x,A}$ and $u_{x,B}$ are the instantaneous local fluctuation velocity components in the x direction at points A and B. The correlation function $R_{A,B}$ is a function of the distance, r_x, between A and B: $R_{AB}(r_x)$.

In the case of uniform velocity and homogeneous turbulence (the fluctuation velocities are small in comparison with the mean velocity) the cross correlation function (47) can be replaced by an autocorrelation function $R_E(\tau)$ (Eq. 48) by means of relationship (49):

$$R_E(\tau) = \frac{\overline{u_{x,A}(t)\, u_{x,A}(t+\tau)}}{u_{x,A}(t)^2} \qquad\qquad (48)$$

$$r_x = \bar{U}_x(t) \qquad\qquad (49)$$

Here $\bar{U}_x$ is the mean velocity in the x direction and
   τ the variable time lag.

$R_E(\tau)$ was directly evaluated from the signal of the constant temperature anemometer by a process computer on-line.

4.3.9 Turbulence Power Spectrum

Another means of describing turbulence structure is the use of spectra. According to this concept, turbulence signals consist of particular overlapping signals of different frequencies and amplitudes. A definite eddy size can be associated with each particular frequency. High frequencies represent small eddies and low frequencies large eddies. Between a one dimensional turbulence power spectrum, $E_1(n)$, and an autocorrelation function, $R_E(\tau)$ a relationship (50) exists [262]:

$$E_1(n) = 4u'^2_x \int_0^\infty R_E(\tau) \cos(2\pi n\tau)\, d\tau \qquad\qquad (50)$$

This power spectrum can be calculated from the autocorrelation function, $R_E(\tau)$, or directly from the original signal by Fourier transform, employing Hanning as well as Parzen windows [263]. Ensemble means were evaluated by accumulation of the individual results.

Both of these methods were used to evaluate $E_1(n)$ or $E_1(k)$. $E_1(k)$ is given by Eq. (51)

$$E_1(k) = \frac{\bar{U}}{2\pi} E_1(n) \tag{51}$$

where $k = \frac{2\pi n}{\bar{U}}$, the wave number and

  $n = $ frequency.

4.3.10 Scales of Turbulence

The intensity of mixing has a close relationship to the macro scale; the degree of gas dispersion to the micro scale. The scales can be evaluated from the cross correlation function or, in homogeneous turbulence, from the autocorrelation function as well as from the turbulence spectrum.

The macro scale Λ_f, is defined by Eq. (52)

$$\Lambda_f = \int_0^\infty R(r_x)\, dr_x \tag{52}$$

and the micro scale λ_f by Eq. (53):

$$\frac{1}{\lambda_f^2} = -\frac{1}{2}\left[\frac{d^2 R(r_x)}{dr_x^2}\right]_{r_x=0} \tag{53}$$

In Eqs. (52) and (53), $R(r_x)$ is the cross correlation function (Eq. (47)).

Because of the difficulties in determining λ_f^2 according to Eq. (53) Taylor [264] recommended the following approximation which is based on a parabolic fitting at $r_x = 0$:

$$R \simeq 1 - \frac{r_x^2}{\lambda_f^2} \tag{54}$$

Microscales can also be determined from the gradient of velocity fluctuations:

$$\frac{1}{\lambda_f^2} \simeq \frac{1}{u_x^2}\overline{\left(\frac{\partial u_x}{\partial r_x}\right)^2} \tag{55}$$

In a steady state system, the square of the turbulence intensity represents the sum of the energy carried by the eddies.

$$u_x'^2 = \int_0^\infty E_1(k)\, dk = \int_0^\infty E_1(n)\, dn \tag{56}$$

Thus macro and micro scales can also be evaluated from power spectrum:

$$\lim_{n \to 0} \frac{1}{u_x'^2} E_1(n) = \frac{4}{\bar{u}} \Lambda_f \tag{57}$$

and

$$\frac{1}{\lambda_f} = \frac{2\pi^2}{\bar{U}^2 u_x'^2} \int_0^\infty n^2 E_1(n) \, dn \tag{58}$$

When employing the autocorrelation function, $R_E(\tau)$ (Eq. (48)), the dissipation time scale, τ_E, can be evaluated from it analogous to Eq. (53), or by Eq. (59):

$$\tau_E = \frac{2\sqrt{u_x^2}}{\sqrt{\left(\frac{\partial u_x'}{\partial t}\right)^2}} \tag{59}$$

τ_E can be used to calculate λ_f by Eq. (60),

$$\lambda_f = \tau_E U \tag{60}$$

if the turbulence has quasi-rigid structure.

4.3.11 Energy Dissipation Spectrum

One can recognize from Eq. (58) that at high frequencies energy dissipation is high. According to Taylor [264], the following relationship holds true for isotropic turbulence:

$$-\frac{du_x'^2}{dt} = 20\nu \frac{u_x'^2}{\lambda_f^2} \tag{61}$$

where ν is the kinematic viscosity.

Equations (58) and (61) yield, for the per unit mass and time dissipation energy, E:

$$E = 15\nu \frac{4\pi^2}{\bar{U}^2} \int_0^\infty n^2 E(n) \, dn \tag{62}$$

These relationships are based on the isotropic turbulence condition.

However, Kolmogoroff [265] developed a general hypothesis for local isotropic turbulence:
At large Reynolds numbers of turbulent motion, the local property of turbulent motion should have a universal character described by the following concepts. First, it is locally isotropic whether the large scale motions are isotropic or not. Second, the motion at very small scales is chiefly governed by viscous forces and the amount of energy which is handed down to them from the larger eddies. In this long series of processes leading to the smallest eddies, the turbulent motion adjusts itself to some definite state. The further down the scale, the less the motion is dependent on the large eddies. Kolmogoroff postulates that all dissipation occurs at the smallest scales when the Reynolds number of turbulent motion are sufficiently high. He has shown that in a dissipation range with a length scale, η^*, and velocity scale, v, Eqs. (63) and (64) are valid:

$$\eta^* = \left(\frac{\nu^2}{E}\right)^{1/4} \tag{63}$$

$$v = (\nu E)^{1/4} \tag{64}$$

For the inertial subrange, the following relationship is valid:

$$E_1(k) = mE^{2/3}k^{-5/3} \tag{65}$$

where m ist a constant.

Similar relationships were also developed by v. Weizsäcker [266] who has shown that slopes of $k^{-5/3}$ correspond to quasi-rigid space structure, whereas a power spectrum slope of k^{-2} indicates substantial structure fluctuations with time. Relationship (65) is also valid for lower Reynolds numbers according to v. Karman and Howard [268]. If is therefore expected that this relationship also holds true for bubble swarm turbulence, which prevails in bubble columns. For the viscous dissipation range, a k^{-7} law holds, which was developed by Heisenberg [267].

Eddies of different sizes have different energy dissipation capabilities. This can be represented by an *energy dissipation spectrum* $\Phi_1(k)$:

$$\Phi_1(k) = k^2 E_1(k) \tag{66}$$

The dissipated energy, E, is a function of the wave number, k:

$$E = 2v \int_0^\infty \Phi_1(k)\,dk \tag{67}$$

From Eqs. (66) and (67) one can recognize that the smallest eddies have the highest energy dissipation rates, whereas their energy content is negligibly small in comparison with the large eddies. The shape of $\Phi_1(k)$ is characteristic for the efficiency of turbulence gas dispersion since only energy, which is dissipated by small eddies can contribute to the gas dispersion.

The energy dissipation spectrum was calculated from the power spectrum $E_1(k)$ according to Eq. (67).

5 Mathematical Models

Mathematical models were employed to describe the growth process, identify the Monod parameters and simulate the reactor. Depending on the reactor type considered, different models can be applied.

5.1 Counter Current Multistage Tower Loop Reactors

Since the liquid mixing time in each stage was much shorter than the mean liquid residence time, it was assumed that each stage can be described by a CSTR model (i.e. in each stage perfect liquid mixing prevails) and neither liquid back flow nor backmixing between the stages exist, since the liquid flows from the stage (n − 1) to the lower stage n through the down-comer tube, which has a much smaller diameter than the column. Only gas passes through the trays so that a cascade model consisting of CSTR's can be employed to describe a multistage tower reactor.

For n-th stage the following quasi-steady state balances are valid:

$$D(X_n - X_{n-1}) = \mu_m X_n \tag{68}$$

for the cell mass balance,

$$D(S_n - S_{n-1}) = -\frac{\mu_m}{Y_{X/S}} X_n \tag{69}$$

for the substrate balance and for the oxygen balances

$$D(O_n - O_{n-1}) = -\frac{\mu_m}{Y_{X/O}} X_n + k_L a (O_n^* - O_n) \tag{70}$$

if nonlimiting growth and a quasi-steady state prevail.

For substrate limited growth, Eq. (69) is used and μ_m was replaced μ_s

$$\mu_s = \mu_m \frac{S}{K_s + S} \tag{71}$$

and for oxygen transfer limited growth, Eq. (70) is used and μ_m was replaced by μ_o

$$\mu_o = \mu_m \frac{O}{K_o + O} \tag{72}$$

The cell concentrations, X_n, substrate concentrations, S_n, dissolved oxygen concentrations, O_n, and dilution rate, D, were measured; saturation concentration O_n^* was calculated from the gas composition and the O_2 solubility in the medium. μ_m, μ_s and μ_o, as well as Y and $Y_{X/O}$ were determined from the corresponding mass balances. $k_L a$ was calculated by Eq. (37).

5.2 Single Stage Tower Loop Reactor

Several papers have considered the mathematical modelling of tower reactors [13, 14, 16, 17, 24, 31, 32, 153, 181, 269–274, 276–280, 282, 285–289]. However, only a few of their results have been applied to biological systems and consider parameter identification and process-simulation, employing real bioreactors which were sufficiently characterized to justify the use of sophisticated models [181, 269–272]. Only these models are considered here, because the evaluation of model parameters, which is discussed in detail, were carried out by identification of these parameters.

During nonlimited and oxygen transfer limited growth the medium can be considered as perfectly mixed with the exception of the dissolved oxygen concentration. In the following, only the oxygen mass balance will be considered for oxygen transfer limited growth as an example. For details of model development, parameter identification and process simulation, see Refs. [181, 269–271]. In the case of substrate-limited growth, the spacial nonuniformity of substrate concentration also has to be taken into account. It is not considered here because of the complexity of this model system. For complete treatment of these systems, see Refs. [118, 272].

5.3 Oxygen Transfer Limited Growth in Batch Operation

The general momentum, heat and mass balance equation system was simplified by the following assumptions: Heat and momentum balances need not be considered.

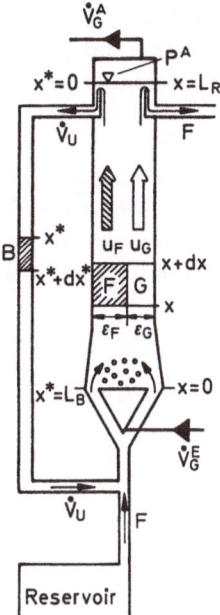

Fig. 1. Schematical view of the mathematical model employed for the single concurrent (air lift) tower loop reactor [181, 269]

The reactor contains only two phases (gas-liquid-system), i.e., microorganisms are not considered as a separate phase. The two-phase flow is homogeneous in the angular and radial directions. For all components in the liquid phase, the same transport velocities and longitudinal dispersion coefficients are valid. The reactor consists of a tower section which is aerated (two phase flow) and a loop which is not aerated (liquid flow alone). In the loop, back mixing is negligible due to high liquid velocity.

According to these assumptions, the general balance equation system reduced to a twin reactor model, which is shown in Fig. 1.

This model is based on mass balances of the following components: cells, substrates, dissolved oxygen, carbon dioxide and nitrogen, in the liquid phases of the tower section and loop, as well as on mass balances of oxygen and carbon dioxide in the gas phase of the tower section. Since except for the liquid velocity, all process variables are angular and radially nearly uniform, their variations are only considered in the longitudinal direction in the tower section and loop. Each of these phases is characterized by a single phase velocity and a longitudinal dispersion coefficient.

In these simplified differential equations the following process parameters appear:

$E_G(t)$ mean gas hold up in the tower section
$E_L(t)$ mean liquid hold up $= 1 - E_G(t)$
$u_F(t)$ liquid phase velocity (mean mass velocity) in the tower section
$u_G(x, t)$ gas phase velocity (mean mole velocity) in the tower section
$u_B(t)$ liquid velocity in the loop
$P(x, t)$ pressure in the tower section
$D_F(t)$ longitudinal liquid dispersion coefficient in the tower section
$D_G(t)$ longitudinal gas dispersion coefficient in the tower section

Since, according to our assumption, $E_F(t)$ does not depend on x, there is a linear pressure drop along the column. This pressure variation also influences the spacial dependence of $u_G(x, t)$. Fig. 2 shows oxygen mass balances in gas and liquid phases of a volume element of the tower section. In contrast to the constant liquid density, ϱ_F, the gas density, ϱ_G, is time and space dependent.

In addition to the assumptions considered before the mass transfer resistance in the gas phase was neglected in comparison with the one in the liquid phase and it was assumed that at the gas/liquid interface equilibrium prevails with regard to the dissolved gases. This yields the following parabolic differential equations for the oxygen mass balance in the liquid phase:

$$\frac{\partial O_F(x, t)}{\partial t} = D_F(t) \frac{\partial^2 O_F(x, t)}{\partial x^2} - u_F(t) \frac{\partial O_F(x, t)}{\partial x} - R_{O_F}(X, S, O, x, t) +$$

$$+ k_L(x, t) \; a(x, t) \left[O_F^*(x, t) - O_F(x, t) \right] \qquad (73)$$

and in the gas phase:

$$P(x, t) \frac{\partial x_{OG}(x, t)}{\partial t} = D_G(t) \frac{\partial}{\partial x} \left(P(x, t) \frac{\partial x_{OG}(x, t)}{\partial x} \right) -$$

$$- \frac{\partial}{\partial x} \left(P(x, t) \, x_{OG}(x, t) \cdot u_G(x, t) - \frac{R_G T \; E_F(t)}{M_{O2} E_G(t)} \right) +$$

$$+ k_L(x, t) \; a(x, t) \left[O_F^*(x, t) - O_F(x, t) \right] \qquad (74)$$

The following initial and boundary conditions are assumed:
At the beginning of cultivation, before the medium is inoculated, the medium is saturated with oxygen

$$O_F(x, o) = O_F^*(x, o) \qquad (75)$$

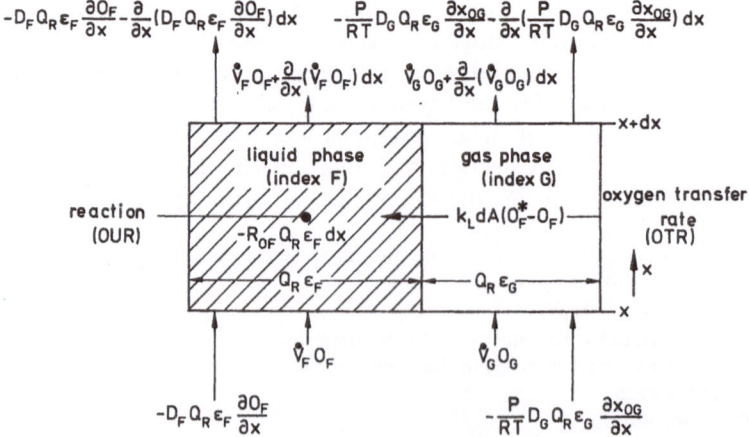

Fig. 2. Mass balances in the gas and liquid phases of a reactor volume element [181, 269]

and the inlet oxygen mole fraction in the gas is preserved along the column

$$x_{OG}(x, 0) = x_{OG}^E \tag{76}$$

During cultivation, Danckwerts boundary conditions are valid at the bottom boundary of the liquid

$$\frac{\partial O_F(0, t)}{\partial x} = \frac{u_F(t)}{D_F(t)} [O_F(0, t) - O_F^E(t)] \tag{77}$$

and in the gas phase:

$$\frac{\partial x_{OG}(0, t)}{\partial x} = \frac{u_G^E(t)}{D_G(t)} [x_{OG}(0, t) - x_{OG}^E] \tag{78}$$

At the top boundary of the liquid

$$\frac{\partial O_F(H, t)}{\partial x} = O \tag{79}$$

and in the gas phase

$$\frac{\partial x_{OG}(H, t)}{\partial x} = O \tag{80}$$

The unknown dissolved oxygen concentration in the liquid phase at the entrance, O_F^E, can only be calculated by means of a loop balance equation (Fig. 3). This equation is given by the following hyperbolic differential equation:

$$\frac{\partial O_B(x^*, t)}{\partial t} = -u_B(t) \frac{\partial O_B(x^*, t)}{\partial x^*} - R_{OB}(X_B, S_B, x^*, t) \tag{81}$$

which is a balance equation for a PFR.

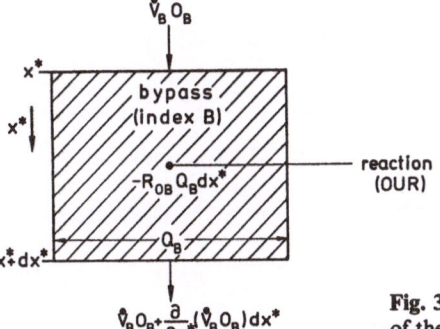

Fig. 3. Oxygen mass balance in the volume element of the loop liquid [181, 269)]

For solution of Eq. (81) the initial condition

$$O_B(x^*, t) = O_F^*(H, 0) \tag{82}$$

and boundary condition at the lower end of the loop

$$O_B(0, t) = O_F(H, t) = O_F^A(t) \tag{83}$$

are used. The latter is controlled by the exit concentration of the tower section, $O_F^A(t)$.

The solution of the oxygen balance equations is only possible if the conditions at the lower end of the tower section are defined by Eq. (84):

$$\frac{\partial O_F(0, t)}{\partial x} = \frac{u_F(t)}{D_F(t)} [O_F(0, t) - \gamma(t) \ O_B^A(t) - \{1 - \gamma(t)\} O_R(t)] \tag{84}$$

where $\gamma(t)$ is the liquid feed back ratio, and the space variant gas velocity, u_G, is known.

When assuming that oxygen and CO_2 exchanges occur simultaneously RQ can be used to determine the CO_2 balance. This allows to calculate the variation of u_G along the reactor (for more details see [181, 269]). In the system considered here the substrate and cell mass balances are relatively simple:

Substrate balance in the tower section:

$$\frac{\partial S_F(x, t)}{\partial t} = D_F(t) \frac{\partial^2 S_F(x, t)}{\partial x^2} - u_F(t) \frac{\partial S_F(x, t)}{\partial x} - R_{SF}(X_F, S_F, O_F, x, t) \tag{85}$$

and in the loop:

$$\frac{\partial S_B(x^*, t)}{\partial t} = -u_B(t) \frac{\partial S_B(x^*, t)}{\partial x^*} - R_{SB}(X_B, S_B, O_B, x^*, t) \tag{86}$$

with the boundary conditions for (85):

$$\frac{\partial S_F(0, t)}{\partial x} = \frac{u_F(t)}{D_F(t)} [S_F(0, t) - \gamma(t) \ S_B^A(t) - \{1 - \gamma(t)\} \ S_R(t)] \tag{87}$$

$$\frac{\partial S_F(H, t)}{\partial x} = O \tag{88}$$

and for (86):
$$S_B(0, t) = S_F^A(t)$$

Initial conditions are given by Eq. (89)

$$S_F(x, o) = S_B(x^*, o) = S_0 \tag{89}$$

where S_0 is the substrate concentration in the substrate reservoir.

The cell mass balance in the liquid phase of the tower section is given by Eq. (90):

$$\frac{\partial X_F}{\partial t} = D_F(t) \frac{\partial^2 X_F(x, t)}{\partial x^2} - u_F(t) \frac{\partial X_F(x, t)}{\partial x} + R_{XF}^*(X_F, S_F, O_F, x, t) \quad (90)$$

and in the loop:

$$\frac{\partial X_B(x^*, t)}{\partial t} = -u_B(t) \frac{\partial X_B(x^*, t)}{\partial x^*} + R_{XB}(X_B, S_B, O_B, x^*, t) \quad (91)$$

Again boundary conditions link these two differential equations:

$$\frac{\partial X_F(0, t)}{\partial x} = \frac{u_F(t)}{D_F(t)} [X_F(0, t) - \gamma(t) X_B^A(t) - \{1 - \gamma(t)\} X_R(t)] \quad (92)$$

$$\frac{\partial X_F(H, t)}{\partial x} = O \quad (93)$$

$$X_B(0, t) = X_F^A(t) \quad (94)$$

For sterile feed $X_R(t) = 0$.

$$R_X^*(\bar{x}, t) = [\mu(S, O, \bar{x}, t) - \mu_T] X(\bar{x}, t) \quad (95)$$

and

$$\mu(S, O, \bar{x}, t) = \mu_m \frac{S(x, t)}{K_S + S(\bar{x}, t)} \frac{O(\bar{x}, t)}{K_o + O(\bar{x}, t)} \quad (96)$$

where u_T is the specific death rate and $\bar{x} = x$ or x^*.

Cell growth rate, R_X, is linked to substrate uptake rate, R_S, by the substrate yield coefficient, $Y_{X/S}(t)$ and to the oxygen uptake rate, R_O, by the oxygen yield coefficient, $Y_{X/O}(t)$.

For more details on these models and others (fed batch, extended culture, continuous models) and for the numerical solution of the necessary differential equation systems, for parameter identification and process simulation carried out on a hybrid computer, see Refs. [181, 269–272, 308].

6 Experimental Systems

6.1 Apparatus

Four different tower loop reactor types were used for the investigations: Two concurrent air lift tower loop systems (a single-stage and a ten-stage reactor) and two counter current tower loop systems (a single-stage and a three-stage reactor with different bubbling layer heights.

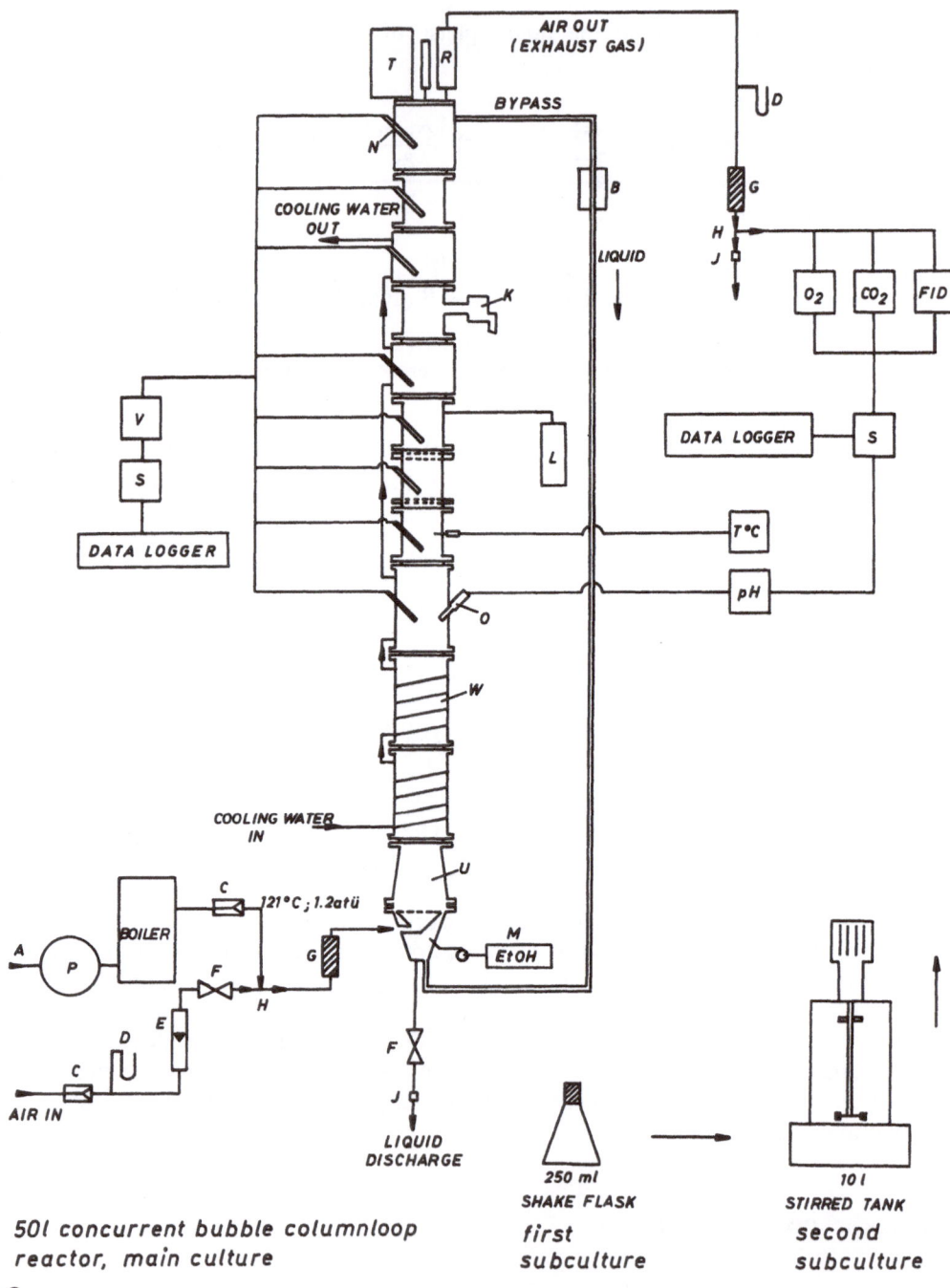

50 l concurrent bubble columnloop reactor, main culture

first subculture

second subculture

a

Fig. 4a. Schematical view of the single-stage concurrent (air lift) tower loop reactor (Reactor A) [16, 153, 154].

A water supply for steam generator; B flow meter; C pressure reducing valve; D pressure gauge; E gas flow meter; F valve; G filter for air sterilization; H three-way-valve; J condensation collector; K sampling; L alkali reservoir; M substrate reservoir; N O_2-electrode; O pH-electrode; P pump; R exhaust gas cooler; S recorder; T engine for mechanical foam destroyer; U porous plate; V amplifier; W heat exchanger; O_2 O_2 gas analyzer; CO_2 CO_2 gas analyzer; pH pH control

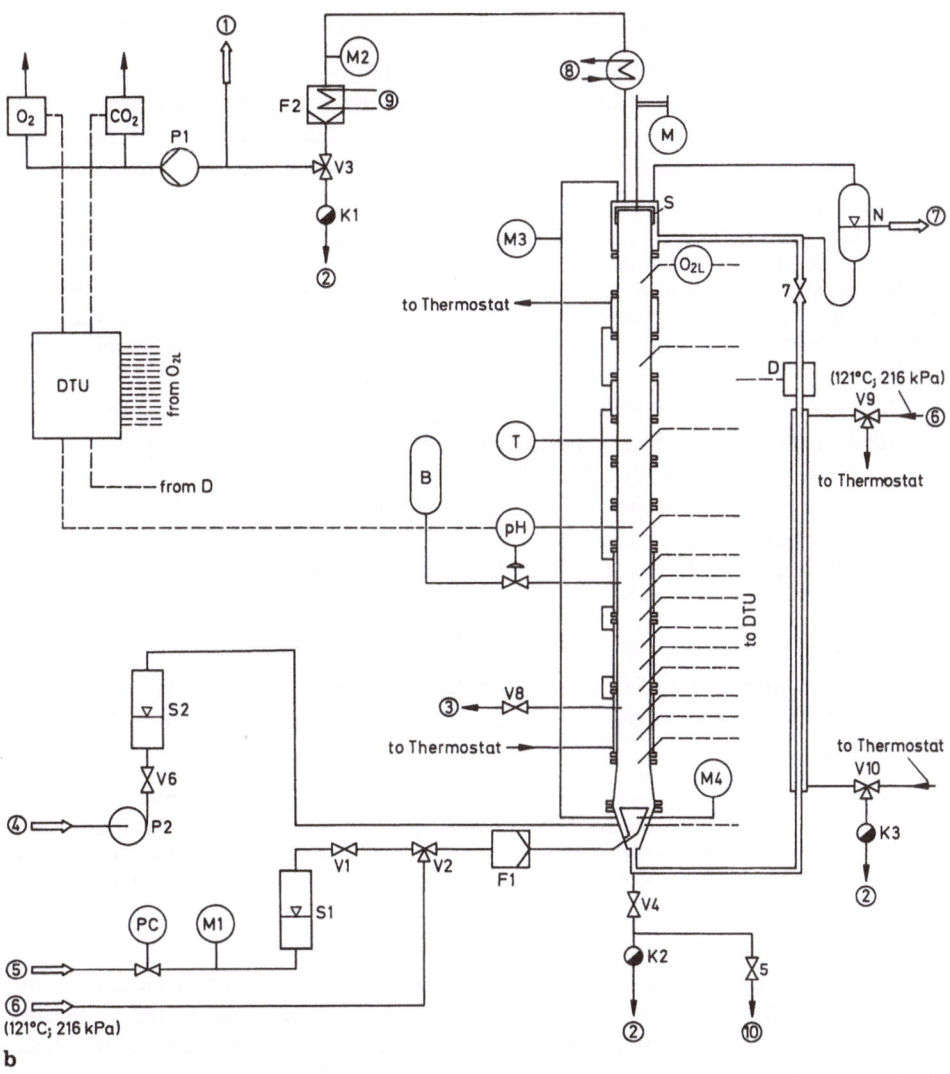

Fig. 4b. Schematical view of the ten-stage concurrent (air lift) tower loop reactor (Reactor B)[17, 178, 305].

PC pressure-reducing valve; M1, M2, M4 pressure gauge; S1, S2 gas flow meter; V1, V6, V7 throttle valves, V2, V3, V9, V10 three/two-way valves; F1 sterile filter; SM mechanical foam destroyer; F2 heated exit gas section; K1—K3 condenser trap; P1 membrane compressor; V4, V5, V8 shut-off valves; P2 centrifugal pump; N level control and overflow; D inductive flow meter; O_{2L} dissolved oxygen measuring electrodes and amplifiers; pH pH meter and control; B supply tank; T temperature meter (Pt 100); M3 difference pressure gauges; O_2 paramagnetic O_2 analyser; CO_2 infrared CO_2 analyser; DTU data transfer and storage unit;
(1) exit gas; (2) condensed water; (3) sampling; (4) feed; (5) air in; (6) steam; (7) medium exit; (8) cooling water; (9) electrical heating; (10) drain.
The instruments for bubble size and turbulence measurements are not given in this figure

6.1.1 Single- and Ten-stage Concurrent Air Lift Tower Loop Reactors

A stainless steel (V4-A) single stage bubble column (air lift) loop reactor, 15 cm in diameter, with a 275 cm high bubbling layer and a stainless steel porous plate (17.5 μm mean pore diameter) was employed for the cultivation of *Hansenula polymorpha* and *Escherichia coli* (Fig. 4). This tower will be termed *Reactor A*.

Air and medium passed the column occurrently and the medium is recirculated through the 2.6 cm inner diameter loop. A mechanical foam destroyer was employed to make it possible to cultivate the microorganisms in the absence of antifoam agents.

The apparatus was sterilized using water vapor (121 °C/2.2 bars). An oil free membrane compressor supplied the air, which was cleaned and sterilized by a filter system. The exhaust gas passed the mechanical foam destroyer, a cooler and a sterile filter, before it left the system. For more details see[16].

The same tower was also used as a ten-stage reactor. Nine perforated plates were installed in the tower, which separated the tower into 10 sections of the following heights (from the top) (Fig. 4a): 1×580 mm, 2×330 mm, 6×200 mm and 1×350 mm. Each of the perforated plates had 163 holes 3 mm in diameter and a relative free cross section of 6.53 %. *E. coli* was cultivated in this reactor, which will be termed *Reactor B*. For more details see Ref. [17].

6.1.2 Single- and Three-stage Countercurrent Tower Loop Reactor

The reactor consists of a 254 cm high stainless steel tower, 20 cm in diameter, which can be operated in a single- or three-stage mode. Fig. 5 shows the three-stage equipment.

At the bottom of the column a perforated plate aerator, 12.4 cm in diameter, with 302 holes, each of them 0.5 mm in diameter, is installed. The tower is equipped with a mechanical foam destroyer at the head and a gas exit with a cooler. The main part of the liquid was pumped by a magnetic rotational pump from the bottom of the column through the 1.7 cm in diameter loop, back to the top of the column.

The single stage tower will be termed *Reactor C*.

Separating trays were employed in the three-stage column at distances 66 and 116 cm from the aerator compartment. They consisted of a perforated plate with 540 holes, 0.5 mm in hole diameter (free cross sectional area 0.34 %) and an overflow, 46 mm in diameter, with 20 and/or 40 cm overflow heights. The stages are numbered from the top to the bottom in direction of the liquid flow. The 40 cm overflow tower will be termed *Reactor D-40*; the 20 cm overflow tower *Reactor D-20*.

An oil-free compressor supplied this equipment with air, which was sterilized by a sterile filter. The exhaust gas left the reactor through a cooler and a sterile trap. For more details see Ref. [87].

Hansenula polymorpha was also cultivated in Reactors C and D.

6.2 Measuring Devices

In order to characterize and control the reactors, several measurements were carried out on the systems and on samples taken from them. These measurements will be divided here into on-line and off-line measurements.

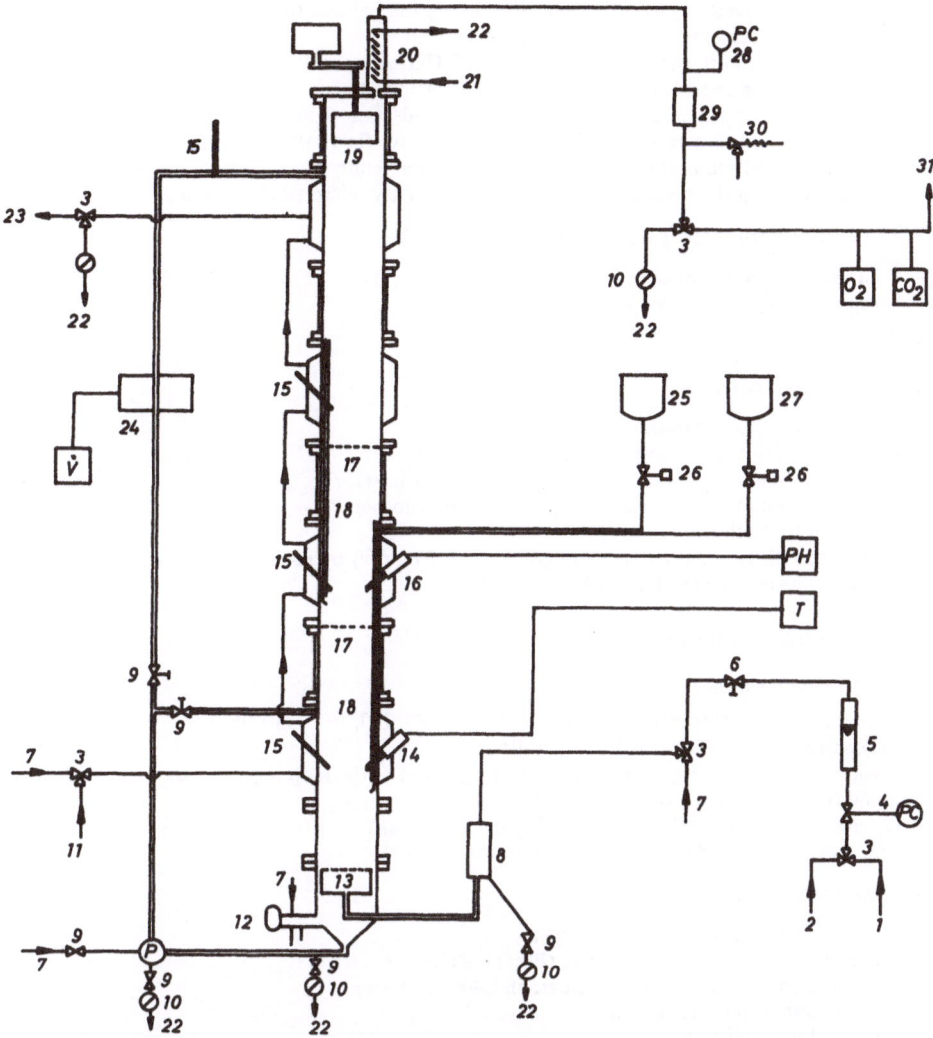

Fig. 5. Schematical view of the three-stage countercurrent tower loop reactor (Reactor D) [87, 88, 89]. *1* air supply; *2* N$_2$ supply; *3* three-way valve; *4* pressure reduction valve; *5* gas flow meter; *6* needle valve; *7* water vapor (121 °C); *8*. sterile filter; *9* valve; *10* condensation collector; *11* thermostatized water inlet; *12* sampling valve; *13* gas distributor; *14* resistance thermometer; *15* O$_2$ electrode; *16* pH electrode; *17* compartment separating tray; *18* overflow tube; *19* mechanical foam destroyer; *20* exhaust gas cooler; *21* cooling water; *22* waste water; *23* thermostatized water backflow; *24* inductive liquid flow meter; *25* alkali reservoir; *26* magnetic valve; *27* substrate reservoir; *28* pressure gauge; *29* heated exhaust gas tube sterile gas outlet; *30* safety valve; *31* exhaust gas; CO$_2$ CO$_2$-analyzer (exhaust gas); O$_2$ O$_2$-analyzer (exhaust gas); T temperature measure and control; P liquid pump; V̇ inductive flow meter

6.2.1 Reactors A and B

On-line measurements

— pH control (Ingold electrode, PD controller)
— temperature control (resistance thermometer PT 100, PD controller)

— tension of dissolved oxygen at 13 positions along the column (by oxygen electrodes developed in the author's laboratory)
— CO_2 concentration in the exhaust gas (Uras 2T Hartmann and Braun)
— O_2 concentration in the exhaust gas (Oxymat 2, Siemens)
— alcohol concentration in the exhaust gas (flame ionisation detector, RS5 Ratfisch)
— volumetric liquid flow rate through the loop (inductive flow meter, Krohne)
— bubble size distribution (by means of 4 electrical conducting microprobes (see 4.3.2)
— local gas hold-up (by means of 4 electrical conductivity micro probes (see 4.3.2)

Additional measurements with E. coli:

— 2 constant temperature anemometers with 2 wedge shaped probes (DISA 55R32), to measure turbulence properties (see 4.3.6 to 4.3.11).

For more details see Ref. [273]

Off-line measurements

— dry biomass (after separation by weight)
— cell concentration by optical density (PMQ2, Zeiss)
— glucose concentration (polarimeter PM241, Perkin Elmer)
— ethanol concentration (gas chromatograph, L400 Siemens)
— volume of feed substrate
— bubble size distribution (flash photography; see Ref. [226]) by means of three windows
— gas flow rate (rotameter, Krohne)

6.2.2 Reactors C and D

On-line measurements

— temperature control (resistance thermometer type PT 100, PD controller)
— pH control (Ingold electrode, PD controller)
— tension of dissolved oxygen in each of the stages and in the loop (again by oxygen electrodes)
— liquid flow rate in the loop (inductive flow meter, Krohne)
— O_2 concentration in the exhaust gas (Oxygor 3, Maihak)
— CO_2 concentration in the exhaust gas (Unor 6, Maihak)

Off-line measurements

— dry biomass (after separation by weight)
— cell concentration by optical density (PMQ2, Zeiss)
— ethanol concentration (gas chromatograph L400, Siemens)
— gas flow rate (rotameter, Krohne)
— volume of feed substrate.

6.3 Subcultures and Operations

Hansenula polymorpha

In Table 7, the composition of the nutrient solution is given for the first (250 ml) and second (10 l) subcultures, as well as for the main cultures (32–50 l). For the glucose runs in the two subcultures and the main culture, 1 % glucose was used.

For ethanol runs in the first subculture, 1 % glycerol in the second subculture, and 1 % ethanol in the main culture provided the energy source.

The cultivations were carried out at 38 °C and pH 5 in batch, fed batch and in extended culture operations. For more details see Refs. [16, 87].

Escherichia coli

The medium composition of the second subculture (10 l) and main culture (50 l) is given in Table 8. Cultivations were carried out at 28 and 30 °C, pH 6.7–6.9, in batch and continuous operation. For more details see Ref. [17].

Table 7. Composition of the nutrient solution [16, 153]

	g l^{-1}		g l^{-1}
$(NH_4)_2SO_4$	5	$MnSO_4 \cdot H_2O$	0.4×10^{-3}
$K_2HPO_4 \cdot 3\ H_2O$	1	$ZnSO_4 \cdot 5\ H_2O$	0.4×10^{-3}
NH_2PO_4	3	$CuSO_4 \cdot 5\ H_2O$	0.04×10^{-3}
$MgSO_4 \cdot 7\ H_2O$	0.5	KI	0.1×10^{-3}
$NaCl$	0.1	$(NH_4)_6Mo_7O_{24} \cdot 4\ H_2O$	0.2×10^{-3}
KCl	0.1	Thiamine	0.004
H_3BO_3	0.5×10^{-3}	Biotin	0.02×10^{-3}
$FeCl_3 \cdot 6\ H_2O\ 9^{-6}$	0.2×10^{-3}		

Table 8. Nutrient medium for cultivation of *E. coli* in stirred tank and tower loop reactors [17, 305]

1.0% glucose
1.0% casein peptone
0.5% meat extract
1.0% yeast extract
0.25% NaCl
pH 6.7–6.9
0.1–0.2% Desmophen 3600 as antifoam agent

6.4 Properties of the Employed Media

Turbidity temperatures, T_T, are mainly influenced by nutrient salts in the cultivation media. Since the nutrient salt composition and salt concentrations vary only slightly, T_T is nearly constant during cultivation:

$$T_T = 53.3\ °C \quad \text{during cultivation of } H.\ polymorpha \text{ on substrate glucose}$$
$$\text{and}$$
$$T_T = 55.9\ °C \quad \text{during cultivation of } E.\ coli$$
$$\text{(For comparison see Table 6)}$$

Oxygen solubilities are given by Bunsen coefficients, α_B

$$\alpha_B = \frac{\text{abs. gas volume under stand. cond. dissolved}}{\text{liquid volume}}$$

The employed salts, glucose, proteins (casein peptone, yeast extract, meat extract) reduce the O_2 solubility. However, since nutrient salt and dissolved protein concentrations vary only slightly during cultivation, the α_B values are nearly constant:

$$\alpha_B \simeq 22 \times 10^{-3} \quad \text{during } H.\ polymorpha \text{ cultivation on substrate glucose,}$$
$$\text{at } T = 38\ °C,$$
$$\alpha_B \simeq 23 \times 10^{-3} \quad \text{during } E.\ coli \text{ cultivation at } T = 28\text{--}30\ °C. \text{ With}$$
$$\text{decreasing glucose concentration (in a batch culture)}$$
$$\alpha_B \text{ increases [201]}$$

For comparison:

$$\alpha_B = 26.68 \times 10^{-3} \text{ in water at } T = 30 \text{ °C}.$$

The *viscosity* of the medium was only slightly higher than that of water due to low cell mass concentrations used during cultivations. Therefore the *diffusivity* of O_2 was also nearly the same as in pure water.

The *surface tension* of cultivation media, σ, varies during measurement due to the denaturation of proteins on the gas/liquid interface. It takes a fairly long time to attain a constant value, which is called the equilibrium surface tension, σ_{eq}. With increasing cultivation time of *Hansenula polymorpha*, t, (in the absence of antifoam agents) σ_{eq} diminishes from 58 mN m^{-1} (Millinewton per meter), at $t = 0$ to 45 mN m^{-1} at $t = 30$ h [210]. The variation of σ with the measuring time, t_M, can be described by a simple relationship [210]:

$$\log 2.3 \, (\log V) = m \log t_M + \log b$$

where $V = \dfrac{\sigma_o - \sigma_{eq}}{\sigma_t - \sigma_{eq}}$

σ is σ_o at $t = 0$
σ is σ_t at t and
m and b are constants which depend on operational conditions.

In nonlimited growth of *H. polymorpha*, m diminishes with increasing t_M from about 1.0 to 0.3—0.4 (in oxygen transfer limited growth) and/or 0.2—0.3 (in substrate limited growth). m is constant in limited growth operations.

In nonlimited growth, the constant b increases with increasing t_M from 3×10^{-4} to 0.2—0.4 (in oxygen transfer limited growth) or to 0.4—0.6 (in substrate limited growth) [210]. This variation can be explained by the change in the protein structure on the gas/liquid interface, which obviously influences the $k_L a$ values (see also 6.5.1.2).

H. polymorpha cultivation media on substrate glucose have about the same coalescence promoting effect as pure water: $m_{corr} \simeq 1.0$.

H. polymorpha media on substrate ethanol are, to a considerable degree, coalescence suppressing. However, because of strong foaming (no antifoam agent was used) it was only possible to determine m_{corr} at low superficial gas velocities:

$$m_{corr} \simeq 1.3 \text{ at } w_{SG} = 2 \text{ cm s}^{-1} \text{ [217, 218]}$$

In the presence of antifoam agents, the coalescence promoting effect is strong. For *H. polymorpha* on substrate ethanol,

$$m_{corr} = 0.66 \text{ at } w_{SG} = 4 \text{ cm s}^{-1} \quad \text{in the presence of silicon}$$
$$\text{oil or Desmophen 3600 [217, 218]}$$

E. coli cannot be cultivated in the absence of an antifoam agent because of uncontrollable foaming. In spite of the presence of antifoam agents, *E. coli* medium has a coalescence suppressing character:

$$m_{corr} = 1.4 \text{ at } w_{SG} = 4 \text{ cm s}^{-1} \quad \text{in the presence of Desmophen 3600}$$

This property is responsible for the unexpectedly small Sauter diameters, d_S, the high specific interfacial areas, a, and k_La values which were measured in *E. coli* media during cultivation (see also (7.2.1)).

Foaminess is not only a function of protein and salt concentrations but also depends on the measuring time, t_{DG}, which is necessary to attain the equilibrium surface tension, σ_{eq} (*H. polymorpha*):

$$\Sigma = 4.23 \times 10^4 \, (1 + 3.85 \times 10^{-5})^{-t_{DG}}$$

where Σ is given in seconds; t_{DG} in minutes.

Thus, if t_{DG} is small (e.g., 1500 min), Σ is large (1000 s), and, conversely for high t_{DG} (4500 min) Σ is low (2 s) [210].

Cell microflotation is influenced by several parameters. However, the most important factors are the foaminess and the maximum possible cell concentration in the foam. If the foaminess is small and the cell concentration in the medium, X_p, is high, the cell concentration in the foam, X_s, corresponds to the constant saturation value. If the foaminess is high and X_p is low, X_s is low as well, and the foam mainly carries aqueous solution.

In batch operation and at a low cell concentration, X_p, the first fraction has the highest (saturation) X_s value.

H. polymorpha can be microflotated directly from its medium, if the medium foaminess is adequate. This is the case in oxygen transfer limited growth. At low medium cell concentrations ($X_p \simeq 2 \, \text{g} \, l^{-1}$) a high concentrating index ($X_s/X_p > 20$) and enrichment index ($X_s/X_R > 500$) can be attained [247]. Here, c_R is the cell concentration in the rest liquid after cell flotation.

Nonflocculating E. coli cells form a stable cell suspension. Nonflocculating yeast cells have very low *sedimentation* rates. *Saccharomyces cerevisiae*, for example, has a typical sedimentation rate of 0.276 cm h^{-1}, calculated from the displacement rate of the clear liquid/turbid liquid interface. The addition of $CaCl_2$, Chitosan, H_3PO_4, H_2SO_4 (pH 3), Dextrane, or DEAE cellulose has only a slight effect on the sedimentation rates. They vary within 0.2 and 0.28 cm h^{-1}. The corresponding half lives, $t_{1/2}$, which are necessary to reduce the cell concentration (extinction) to half of the original value, halfway up the cylinder, vary between 42 and 53 h. With 0.2% CMC (Tylose C 300, Hoechst AG) the sedimentation rate increases to 2.11 cm h^{-1} and $t_{1/2}$ diminishes to 11 h [256]. Flocculating strains of *E. coli* or *S. uvarum* have high sedimentation rates which cannot be measured using these batch methods but only by continuous methods. They are two to three orders of magnitude higher than the sedimentation rates of nonflocculating cells.

6.5 Properties of the Employed Two-phase Systems

6.5.1 Gas Hold-up, E_G, Bubble Diameters, $\bar{d}$ and d_S, Specific Surface Area, a, and Volumetric Mass Transfer Coefficient, k_La

6.5.1.1 Spacial Variation

The measurements of the local properties of two phase systems indicate that radial profiles of $\bar{d}$ and d_S in tower reactors are fairly uniform [36, 178, 226, 227, 273, 274]. Also their longitudinal variations are fairly moderate except in the neighbourhood of the aerator [178, 274]. The same holds true for the spacial variations of the local E_G's. According to Eq. (35), a is proportional to E_G, $(1 - E_G)^{-1}$ and d_s^{-1}, thus the spacial dependence of E_G is amplified in a. Especially at high superficial gas velocities, i.e. in the heterogeneous flow range, the radial profile of a has a shape of an error function, with its maximum in the column center [36, 226, 227]. Radial profiles of local $k_L a$ values have not yet been published. However, one can assume that they are similar to the radial profiles of a, since there is no reason to assume that k_L is not constant in the tower cross section.

The behaviour of these parameters near the aerator depends on the aerator itself and on the medium character.

If the primary bubble diameter, d_p, is larger than the corresponding dynamic equilibrium bubble diameter, d_e, or if they are equal, $\bar{d}$ and d_s are also constant in the aerator vicinity. Only if $d_p \ll d_e$, does the bubble size increase with increasing distance from the aerator, x, due to coalescence, until d_e is attained. How quickly $d_s \simeq d_e$ is attained depends on the medium property. For uniform longitudinal $\bar{d}$ and d_s-profiles, the local gas hold-up, E_G, is also constant and does not depend on the longitudinal position. This holds true also for the aerator vicinity.

In systems with $d_p \ll d_e$, i.e. in which d_s increases with increasing distance from the aerator, x, the local ε_G passes a maximum and then it diminishes with increasing x. In these systems a depends considerably on the longitudinal position, x, as can be seen from Fig. 6. a attains a constant value as soon as $d_s \simeq d_e$ has been established. Also $k_L a$ exhibits a strong dependence on x in these systems in the vicinity of the aerator [80-87, 275]. Looking for the cause of this dependence, one can recall that in systems with $d_p \ll d_e$, aerators have considerable influence on a, which depends on the medium significantly. In Fig. 7, a^* values are plotted as a function of the superficial gas velocity with three different aerators (different d_p's), in coalescence suppressing medium. The large aerator effect diminishes, as a medium with slight coalescence suppressing effect is employed (Fig. 8).

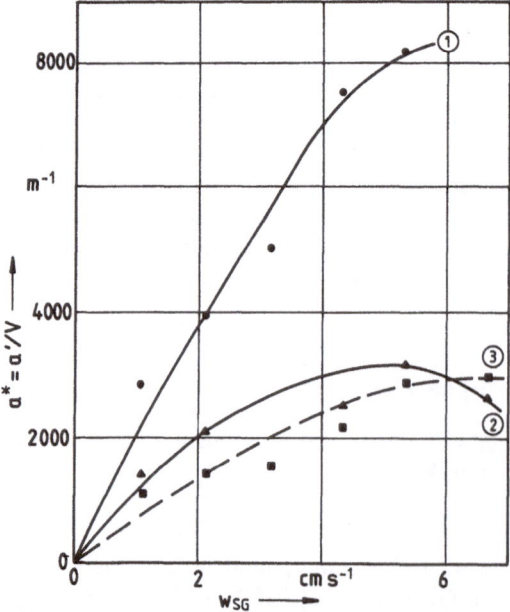

Fig. 6. Specific interfacial area, $a^* = A'/V$, as a function of the superficial gas velocity in a 258 cm higher tower reactor 14 cm in diameter by employing 1% ethanol substrate and nutrient solution according to Table 7. Influence of the longitudinal position, x; 1 distance from the aerator, $x = 41$ cm; 2 $x = 115$ cm; 3 $x = 171$ cm

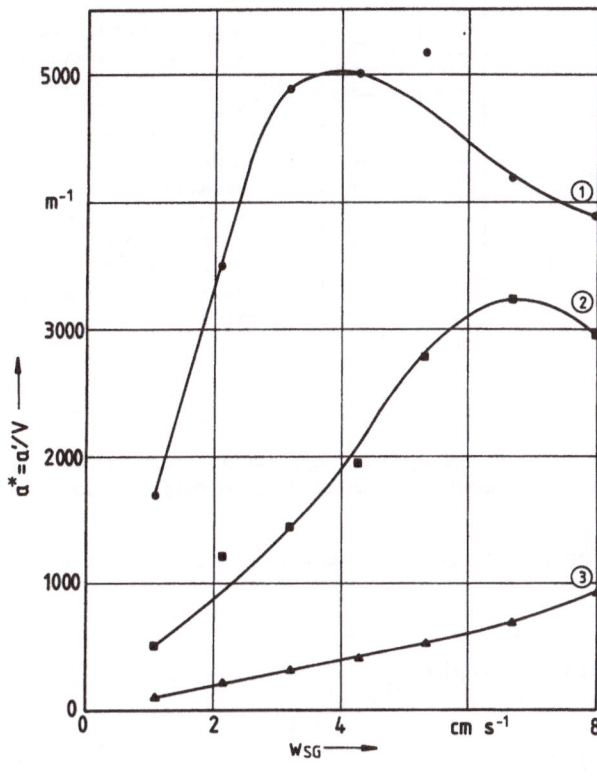

Fig. 7. Specific interfacial area, $a^* = A'/V$, as a function of the superficial gas velocity in a 258 cm high tower reactor 14 cm in diameter by employing 0.5% propanol solution. Influence of the aerator. x = 171 cm; *1* porous plate 5 μm in pore diameter; *2* porous plate 50 μm in pore diameter; *3* perforated plate 500 μm in hole diameter

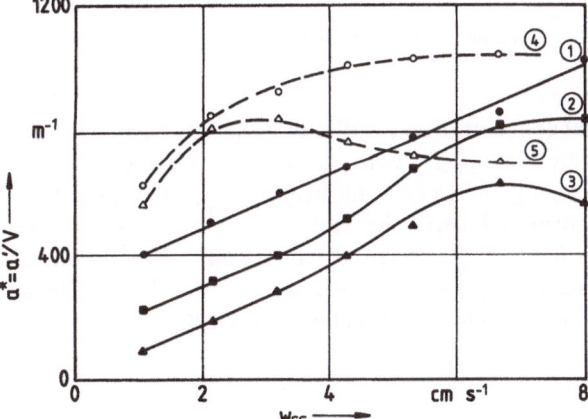

Fig. 8. Specific interfacial area, $a^* = A'/V$, as a function of the superficial gas velocity in a 258 cm high tower reactor 14 cm in diameter by employing 0.5% methanol solution. Influence of the aerator. x = 171 cm; *1* porous plate 5 μm in pore diameter; *2* porous plate 50 μm in pore diameter; *3* perforated plate 500 μm in hole diameter; *4* ejector nozzle 3 mm in diameter; *5* injector nozzle 4 mm in diameter

In coalescence promoting media, this effect is only slight (Fig. 9). This aerator effect on a is well documented in the literature [14, 37, 80–87, 277, 278]. This effect can be taken into account by assuming a variation of $k_L a$ with x near the aerator [16, 153, 181, 279]. However, the variation of $k_L a$ in the vicinity of the aerator is not only due to the space dependence of a, but also to that of k_L [276]. Near the aerator the turbulence intensity is high, which causes a redistribution of the gas phase. With increasing superficial gas velocity, the specific power input increases which reduces d_p [280]. During bubble redistribution, the bubble surface is renewed and a strong "deformation turbulence" due to considerable oscillation and shape variation prevails [281]. All of these phenomena increase

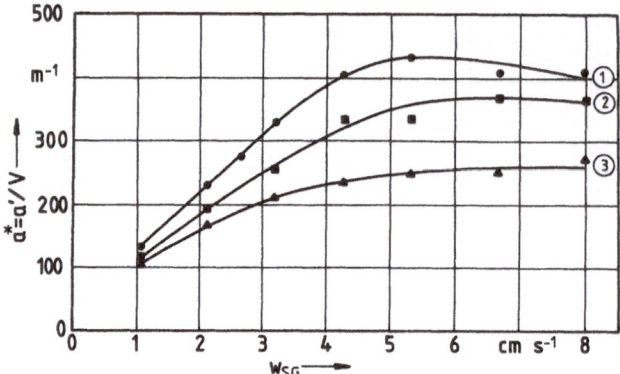

Fig. 9. Specific interfacial area, $a^* = A'/V$, as a function of the superficial gas velocity in a 258 m high tower reactor 14 cm in diameter by employing distilled water. Influence of the aerator. x = 171 cm; *1* porous plate 5 μm in pore diameter; *2* porous plate 50 μm in pore diameter; *3* perforated plate 500 μm in hole diameter

k_L. These spacial dependencies near the aerator of k_L, a and k_La must be taken into account, if one desires to describe longitudinal concentration profiles of dissolved oxygen in the tower reactor.

In the tower loop reactors considered here, this spacial dependence of k_La was calculated by Eq. (97), [181, 282]:

$$k_La(x, t) = \begin{cases} k_La^E(t) \exp\left[-K_{ST}(t) \dfrac{x}{\alpha L_R}\right] & (97\,a) \\ \qquad\qquad\qquad\qquad 0 \leq x \leq \alpha L_R \\ k_La^E(t) \exp\left[-K_{ST}(t)\right] & (97\,b) \\ \qquad\qquad\qquad\qquad \alpha L_R \leq x \leq L_R \end{cases}$$

where k_La^E is the volumetric mass transfer coefficient at the gas entrance (x = 0) and

K_{ST} the coalescence factor

The simulations of longitudinal concentration profiles of dissolved oxygen indicate that all profiles could be fitted by assuming a constant validity range of Eq. (97): $\alpha = 0.1 = z$, where $z = x/L_R$ and $L_R = 276$ cm, the height of the bubbling layer. Thus, in the tower reactors considered here, the variation of k_La is only taken into account in the range x = 0 to x = 27.6 cm. In this range, k_La diminishes exponentially with K_{ST} according to Eq. (97a). In the range x = 27.6 to 276 cm, k_La is constant and given by Eq. (97b). This k_La is called k_La^α here:

$$k_La^\alpha(t) = k_La^E(t) \exp\left[-K_{ST}(t)\right] \qquad (97\,c)$$

The reduction of k_La^E at the reactor entrance to its spacial independent value, k_La^α, is characterized by the coalescence function, Ψ^α, [181, 282]:

$$\Psi^\alpha(t_i) = \frac{k_La^\alpha(t_i)}{k_La^E(t_i)} \qquad (98)$$

which is considerably influenced by the medium properties and operation conditions. For example in the nonlimited growth range of *Hansenula polymorpha* cultivation

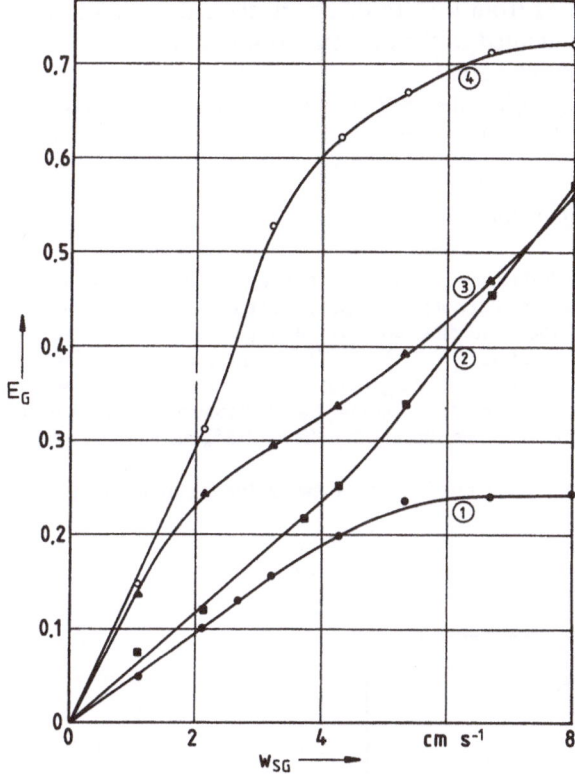

Fig. 10. Mean relative gas hold up, E_G, as a function of the superficial gas velocity in a 258 cm high tower reactor 14 cm in diameter by using porous plate 5 μm in pore diameter. Influence of the medium properties. x = 171 cm; *1* distilled water; *2* 0.5% methanol solution; *3* 0.5% ethanol solution; *4* 0.5% propanol solution

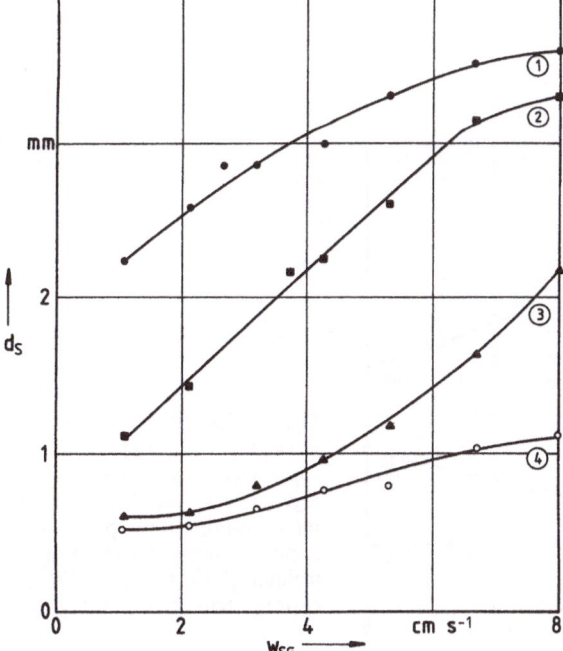

Fig. 11. Sauter bubble diameter, d_S, as a function of the superficial gas velocity in a 258 cm high tower reactor 14 cm in diameter by using porous plate 5 μm in pore diameter. Influence of the medium properties. x = 171 cm (For symbols see Fig. 10)

on substrate ethanol $\Psi^\alpha(t)$ increases from 0.37 to 0.72 with the cultivation time, t. At growth transition from the nonlimited to the oxygen transfer limited state, $\Psi^\alpha(t_i)$ diminishes and then fluctuates around 0.50.

When employing glucose as a substrate, $\Psi^\alpha(t_i)$ has a much lower value. It fluctuates around 0.16 during the first 6 h [181, 282].

6.5.1.2 Influence of Medium Composition

The composition of the medium only influences E_G, $\bar{d}$, d_S, a and k_La, if $d_p \ll d_e$, or if d_e is space dependent. In single-stage tower reactors d_e is not space dependent, except in the aerator vicinity, thus in the predominant section of the tower, a medium effect is only expected if $d_p \ll d_e$. Medium effects are well documented in these reactors [13, 37, 80–87, 148, 277, 278]. In Figs. 10, 11 and 12, E_G, d_S and a^* are plotted as functions of the superficial gas velocity w_{SG}, in a tower reactor with porous plate gas distributor (5 μm mean pore diameter) employing coalescence promoting media (dist. water) and media with increasing coalescence suppressing effects (0.5 % methanol → ethanol → propanol solutions).

However, it is difficult to find quantitative relationships between these parameters and the composition of cultivation media.

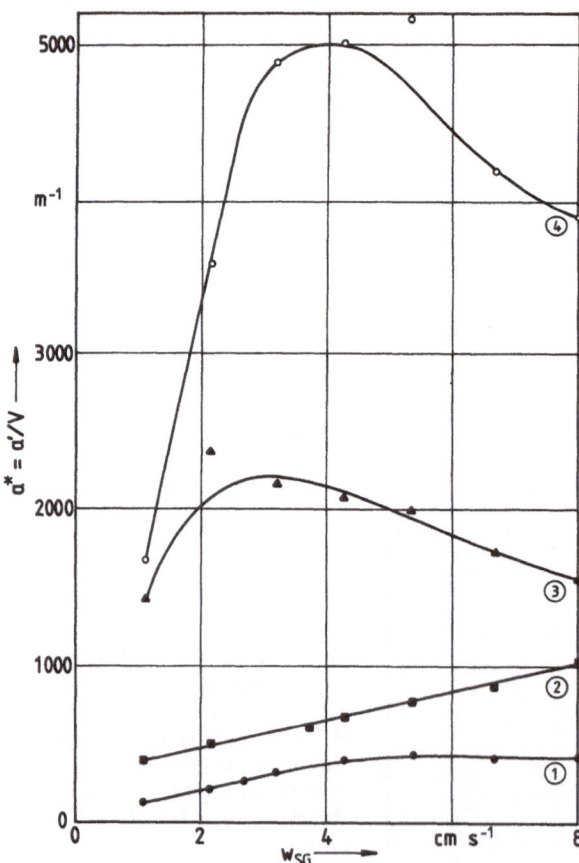

Fig. 12. Specific interfacial area, $a^* = A'/V$, as a function of the superficial gas velocity in a 258 cm high tower reactor 14 cm in diameter by using porous plate 5 μm in pore diameter. Influence of the medium properties. x = 171 cm (For symbols see Fig. 10)

In Fig. 13, identified $k_L a^\alpha$ values are plotted as a function of ethanol concentration. These $k_L a^\alpha$'s where evaluated from different cultivations of *Hansenula polymorpha* with ethanol substrate at the same superficial gas velocity ($w_{SG}^E = 1.9 \text{ cm s}^{-1}$). Identified $k_L a$ values are also plotted for comparison. They were measured by Oels [13, 283] in nutrient salt solutions at S = 0 and S = 5 g l^{-1} ethanol concentrations with the same aerator and at the same superficial gas velocity. With increasing ethanol concentrations, $k_L a^\alpha$ becomes larger. However, $k_L a^\alpha$ seems to depend also on the cultivation mode. At the same substrate concentration, the lowest $k_L a^\alpha$ values were found in oxygen transfer limited cultivations, probably due to higher dissolved protein concentration in the medium caused by higher cell death rate. The lower boundary of $k_L a^\alpha$ values for oxygen limited growth can be represented by a simple straight line. At S = 0 g l^{-1}, $k_L a^\alpha = k_L a_0^\alpha$ and at $S_D = 1.25 \text{ g l}^{-1}$, $k_L a^\alpha = 2k_L a_0^\alpha$ (S_D is defined as the substrate concentration at which $k_L a^\alpha$ attains $2k_L a_0^\alpha$).

In nonlimited growth $S_D = 1 \text{ g l}^{-1}$. For S $\geq 3 \text{ g l}^{-1}$ the dependence of m_L on S can no longer be described by a linear relationship. This relationship holds true for antifoam free systems.

Antifoam additives strongly influence $k_L a$ in tower reactors [14, 217, 218, 284]. However, at low concentrations their concentration also plays a role. When employing antifoam control, the antifoam concentration in cultivation medium is usually unknown. At intermediate concentrations of antifoam additives, $k_L a$ does not depend on their concentrations. High concentrations of antifoam agents can cause foam formation and an increase in $k_L a$.

Because of the complex dependence of $k_L a$ on antifoam type, concentration and medium composition, no relationship is known to calculate this interdependence. However, when employing antifoam agents at intermediate concentrations, the calculation of $k_L a$ appears to be simple.

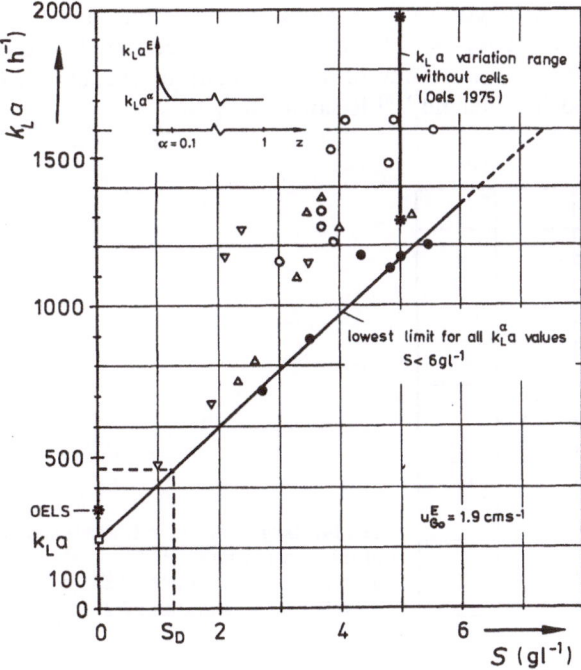

Fig. 13. Influence of ethanol concentration on the volumetric mass transfer coefficient, $k_L a$, during cultivation of *H. polymorpha* [181, 271]. $k_L a_0^\alpha = 235 \text{ h}^{-1}$, $S_D = 1.25 \text{ g l}^{-1}$.
∇△ batch cultivation
□ substrate limited cultivation
○ extended culture
● oxygen transfer limited cultivation
lowest limit for all $k_L a$ values
⊥ $k_L a$ variation range without cells [283]

All cultivations investigated in the author's laboratory gave the same k_La value if an antifoam agent and its concentration were identical [218].

Since it is difficult to remove antifoam agents completely from the inner surfaces of bioreactors, investigations with them are quite tedious and time consuming, which is probably the reason why such investigations are scarce.

The few data published in the literature [217, 218] seem to indicate that the addition of antifoam agents does not influence E_G, but strongly reduces k_La.

The reduction of k_L can be determined at low w_{SG}, where no coalescence occurs. At short bubbling layer heights, the diminution of k_La is mainly caused by the reduction of k_L [217]. Only in higher columns does the increase in d_S have to be taken into account.

6.5.1.3 Influence of Aeration Rate

Aeration rate dependence of E_G, $\bar{d}$, d_S, a and k_La in tower reactors is well documented in the literature (e.g. [13, 30−37, 80−87, 276−280].

In tower reactors with an outer loop, two effects overlap: With increasing gas velocity, E_G, a and k_La are enlarged in the homogeneous flow range similar to tower reactors ("tower effect"). However, in tower loop reactors increasing gas velocity also enlarges the liquid velocity which causes a decrease of the above mentioned parameters ("loop effect").

At low gas velocities, the "tower effect" dominates and at high gas velocities the "loop effect". Hence, with increasing gas velocity in tower loop reactors, k_La at first increases, often passes a maximum and then diminishes [16, 17, 148, 155]. In Fig. 14 OTR is plotted as a function of aeration rate in a tower loop reactor during cultivation of *Hansenula polymorpha*. At 20% saturation of dissolved oxygen concentration, OTR passes a flat maximum as a function of the aeration rate. This maximum occurs at a superficial gas velocity of $w_{SG} = 2.16$ cm s^{-1}. The agreement between the superficial gas velocities at this maximum of OTR found in the author's laboratory [17, 155] and determined by Weiland [148] is satisfactory.

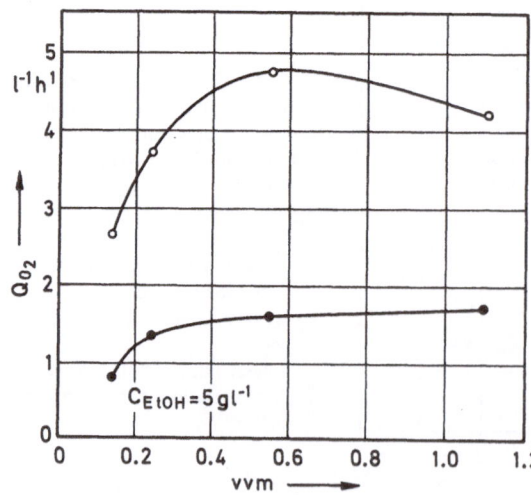

Fig. 14. Oxygen transfer rate as a function of the aeration rate [16, 155].
- • relative O_2-saturation: 0.80
- ○ relative O_2-saturation: 0.20

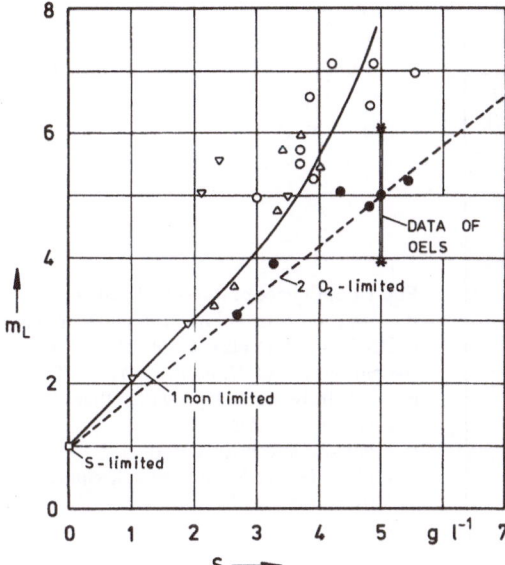

Fig. 15. Mass transfer coefficient ratio, m_L, as a function of the substrate (ethanol) concentration during the cultivation of *H. polymorpha* [181, 282]

1 nonlimited and substrate limited growth
2 lowest limit for all $k_L a$ values (O_2 limited growth)

$\updownarrow$ m_L variation range without cells [283]

$S_D = 1.25 \text{ g l}^{-1}$, $u_{GO}^E = 1.9 \text{ cm s}^{-1}$

At low superficial gas velocities and liquid circulation rates the "tower effect" dominates. In this range the influence of w_{SG}^E on $k_L a^\alpha$ can be taken into account by Eq. (99), by separating the substrate effect from the gas velocity effect [181, 282]:

$$k_L a^\alpha(w_{SG}^E, S) = k_L a_0^\alpha(w_{SG}) \, m_L(S) \qquad (99)$$

where $m_L(S)$ is a coalescence factor ($-$).

Since common nutrient salt solutions have similar coalescence suppressing effects, this effect was integrated into $k_L a_0^\alpha$ of Eq. (99). With $m_L(S)$ the coalescence active substrate concentration was taken into account. Fig. 15 shows m_L as a function of S for different cultivations.

The relationship between m_L and S can be described by Eq. (100):

$$m_L(S) = 1 + \frac{S}{S_D} \qquad (100)$$

where

m_L $(-)$
S (gl^{-1})
$S_D = 1.25$ for oxygen transfer limited growth (curve 2)
$\qquad S \leqq 6 \text{ gl}^{-1}$
$S_D = 1.0$ for nonlimited growth (curve 1) $S \leq 4 \text{ gl}^{-1}$

In Fig. 16, $k_L a_0^\alpha$ is plotted as a function of the superficial gas velocity at the reactor entrance, w_{SG}^E, for cultivations with ethanol and glucose as well as for nutrient salt solutions. This figure indicates again that a low w_{SG}^E values, the agreement between $k_L a_0^\alpha$ in the tower (model media) and in the tower loop (cultivation media) are satisfactory. However, with increasing gas velocity, the "loop effect"

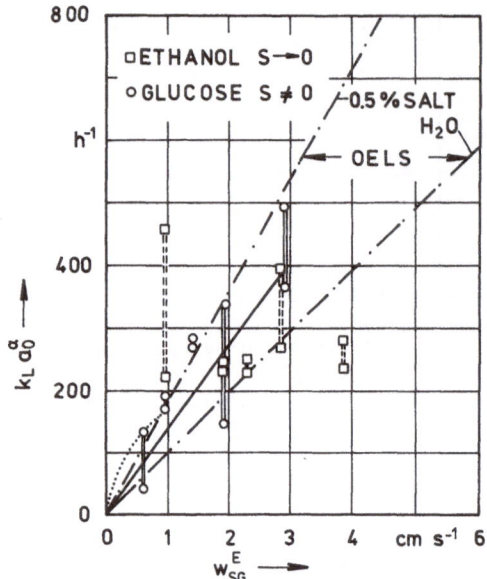

Fig. 16. Influence of the superficial gas velocity, u_{Go}^E, on the volumetric mass transfer coefficient in the absence of ethanol during the cultivation of *H. polymorpha* [181, 282].

□   substrate limited growth, ethanol substrate, $S = 0 \text{ g l}^{-1}$

○   glucose substrate $S \neq 0 \text{ g l}^{-1}$

----   range for 0.5% nutrient salt solution in tower reactor [283]

increases and at high w_{SG}^E it dominates. Thus with increasing w_{SG}^E, $k_L a^\alpha$ passes a maximum between 2 and 3 cm s^{-1}, in contrast to tower reactors in which no such maximum exists. There is a fairly large scattering of the $k_L a_0^\alpha$ data due to the complexity of the biological media. However, there is a clear tendency for $w_{SG}^E \leqq 3 \text{ cm s}^{-1}$ i.e. $k_L a_0$ increases with w_{SG}^E. This can be approximated by Eq. (101):

$$k_L a_0^\alpha = 133.3 \ w_{SG}^E; \qquad w_{SG} \leqq 3 \text{ cm s}^{-1} \qquad\qquad (101)$$

where $k_L a_0^\alpha$ (h^{-1}) and w_{SG}^E (cm s^{-1}).

For $w_{SG}^E > 3 \text{ cm s}^{-1}$, $k_L a_0^\alpha$ diminishes with increasing w_{SG}^E. Relationships (100) and (101) are only valid for antifoam free systems.

6.5.1.4 Influence of Cultivation Time and Operational Mode

The two phase system properties depend considerably on the cultivation time and operation conditions.

Immediately after inoculation, gas hold-up E_G, diminishes, passes a minimum and then gradually and slightly increases. The behaviour of $\bar{d}$ and d_S, resp., corresponds only partly to a mirror image of the E_G course: $\bar{d}$ and d_S increase at first, pass a maximum, a minimum and gradually and slightly increase (Fig. 17). The specific surface area, a, and volumetric mass transfer coefficient, $k_L a^\alpha$, have similar courses at the beginning of the cultivation: both of them diminish after the inoculation, pass a minimum and a maximum. After this maximum, a diminishes only slightly in contrast to $k_L a^\alpha$, which is considerably reduced as cultivation continues.

Similar behaviour was also found with *Candida boidinii* on methanol, ethanol and glucose substrate [13]. This course seems to be characteristic for antifoam free systems. The diminutions of E_G, a and $k_L a$ a after inoculation are due to the increase

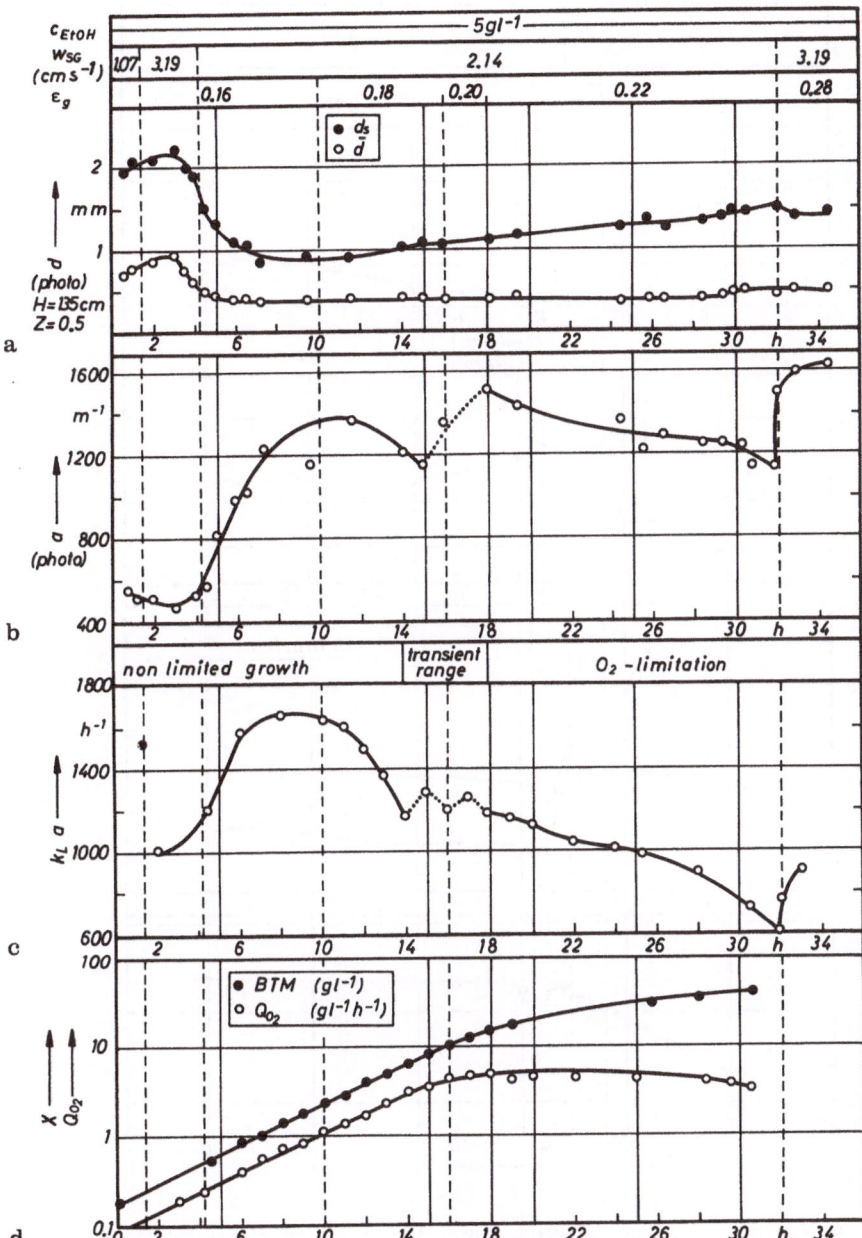

Fig. 17a—d. Variation of system properties during the growth of *H. polymorpha* on ethanol substrate [273, 274].
Extended culture, c_{EtOH} = 5 g l^{-1}, w_{SG} and E_G were varied: **a** mean and Sauter bubble diameter, $\bar{d}$ and d_s; **b** specific interfacial area, a; **c** volumetric mass transfer coefficient, $k_L a$; **d** cell mass, X, and oxygen transfer rate, Q_{O_2}, as a function of the cultivation time

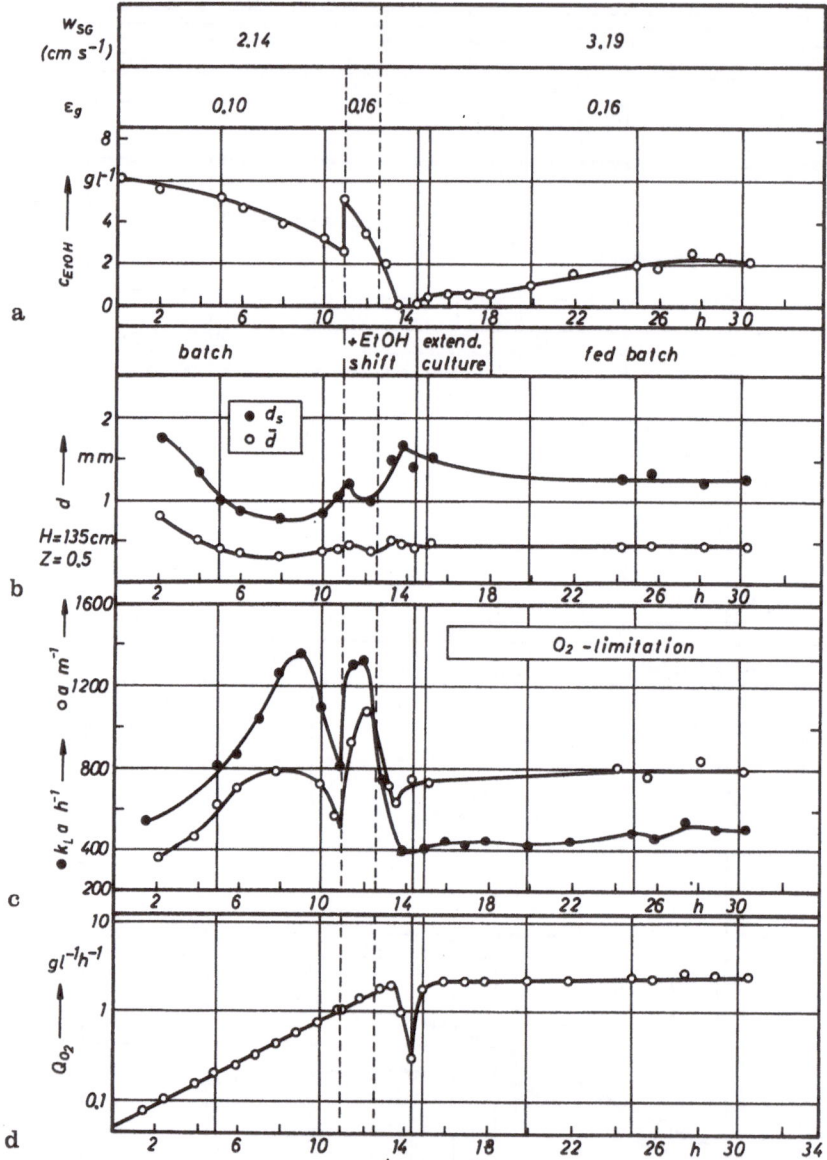

Fig. 18a—d. Variation of system properties during the growth of *H. polymorpha* on ethanol substrate [273, 274]. After batch operation substrate shift was first carried out then extended culture operation with oxygen transfer limitation. w_{SG} and E_G were varied; **a** substrate concentration, C_{EtOH}; **b** mean and Sauter bubble diameter, $\bar{d}$ and d_S; **c** volumetric mass transfer coefficient, k_La; **d** oxygen transfer rate, Q_{O_2} as a function of cultivation time, t

of bubble size caused by considerable coalescence promotion. The origin of this phenomenon is not yet known. This coalescence promotion effect disappears after about 5 h, hence one can assume that it is caused by the yeast cells. They probably secrete surface active components to control their environment: At first to reduce the oxygen tension but after their adaptation to the new system, they consume them to increase the oxygen tension and OTR.

The divergence between the courses of a and $k_L a^\alpha$, after they passed the maximum, can be caused by different phenomena:

— by the diminution of k_L caused by the coverage of the bubble surface by natural surface active substances, which are present in the cultivation media,
— enrichment of small bubbles in the two-phase system.

The gradual increase of E_G with cultivation time seems to indicate that an enrichment of small bubbles exists.

Because of the long small bubble retention time in the column, their oxygen content is totally exhausted. Thus, they no longer contribute to $k_L a$. Hence $k_L a$ is reduced with increasing small bubble hold-up.

The coverage of the interface by surface active substances (fatty acids, denaturated proteins) can only play a significant role for small bubbles with long retention times, since their enrichment at the interface needs considerable time. Since an interface covered by denaturated protein has high surface viscosity and viscoelasticity [213, 214]

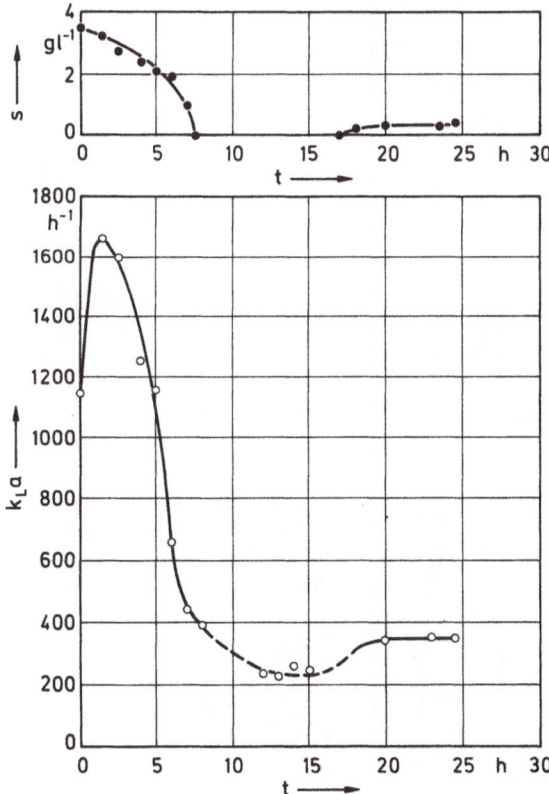

Fig. 19. Substrate concentration, S, and volumetric mass transfer coefficient, $k_L a$, as a function of the cultivation time, t, employing substrate ethanol [16, 282]

● S; ○ $k_L a$

the coalescence of these bubbles is strongly inhibited. Therefore, one can presume that the formation of small bubbles and their surface coverage mutually intensify each other. Thus, the divergence between a and $k_L a$, after they surpassed their maximum, is caused by small bubble formation and enrichment which is accelerated by their surface coverage, but not by the surface coverage of the medium and large bubbles.

Both $k_L a^\alpha$ and a are strongly influenced by the concentration of substrates, which have coalescence suppressing characteristics. This substrate effect is adequately demonstrated in several publications (see 6.5.1.2). From Fig. 18 this effect can clearly be recognized. In the first phase of the cultivation, $k_L a^\alpha$ and a increase, as has already been discussed. At constant and high ethanol concentration a remains high (>1200 m^{-1}) as well as $k_L a^\alpha$ (>1000 h^{-1}) up to t = 25 h (Fig. 17). A reduction of the ethanol concentration, C_{EtOH}, for a short period of time, causes a rapid diminution of a and $k_L a^\alpha$ to fairly low values ($a \simeq 600$ h^{-1}, $k_L a^\alpha \simeq 400$ h^{-1}). If the substrate limitation prevails for longer time, a and $k_L a^\alpha$ are reduced even further, e.g. to $k_L a^\alpha \simeq 200$ h^{-1} (Fig. 19).

6.5.2 Dependence of Turbulent Properties on Space and Operating Parameters

In tower reactors the local mean liquid velocity profiles, $\bar{U}$, are considerably nonuniform. $\bar{U}$ reaches high positive values in the column center and high

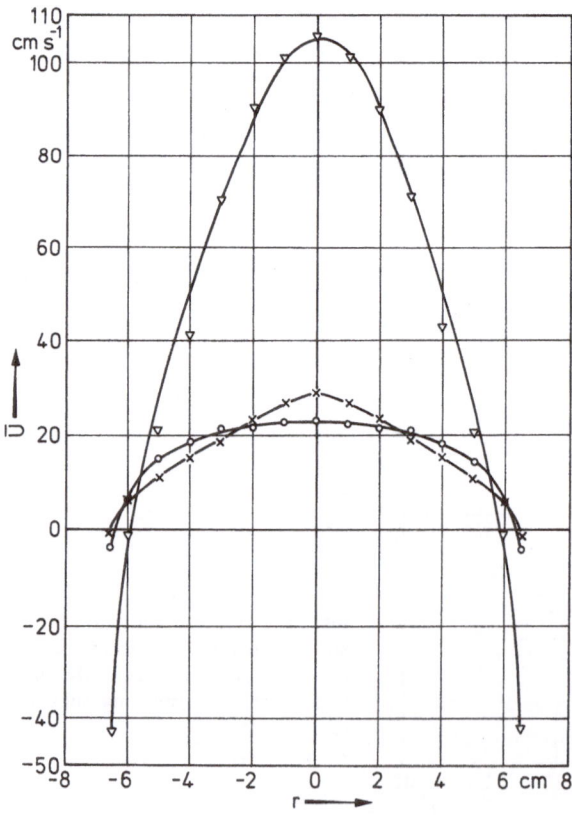

Fig. 20. True mean liquid velocity profiles, $\bar{U}$, in tower reactors using porous plate 5 μm in pore diameter, and 1% methanol solution, X = = 149 cm, w_{SL} = 1,8 cm s^{-1} at different superficial gas velocities, w_{SG} [261].

× w_{SG} = 2.67 cm s^{-1}
○ w_{SG} = 5.35 cm s^{-1}
▽ w_{SG} = 8.02 cm s^{-1}

negative values in the wall range. These profiles are influenced by the aerator type, distance from the aeration rate, w_{SG}, as well as by the medium composition.

At large distances from the aerator, the local mean velocity profile does not depend on x. When using porous plate gas distributors, for which $d_p \ll d_e$, the local mean velocity profiles are flat in the homogeneous flow range (<6 cm s^{-1}) and become considerably nonuniform in the heterogeneous flow range (>6 cm s^{-1}) (Fig. 20). With increasing hole diameter of perforated plate aerator (increasing d_p) these differences gradually diminish.

When using a perforated plate 1 mm in hole diameter the differences between liquid velocity profiles in homogeneous and heterogeneous flow ranges are less, but still significant (Fig. 21) and with a perforated plate 3 mm in hole diameter, these differences completely disappear (Fig. 22). In the last system $d_p \geq d_e$, thus no coalescence is possible and the size of the bubbles is nearly uniform. Therefore, as long as the bubbles can carry the gas which was fed into the system, homogeneous flow prevails. Only at high gas flow rates at which the bubble velocities are not high enough to carry the gas amount, bubble aggregates are formed which rise

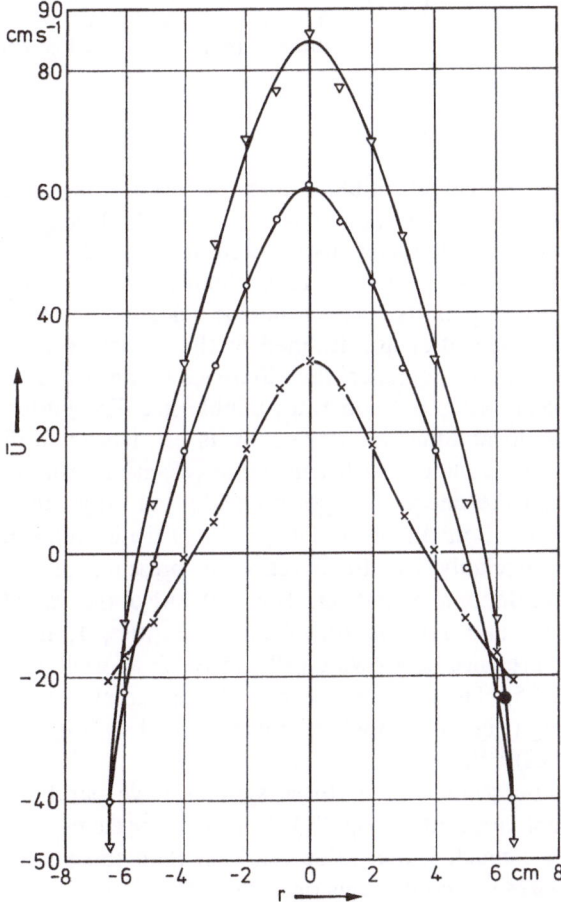

Fig. 21. True mean liquid velocity profiles, Ǔ, in tower reactors using perforated plate 1.0 mm in hole diameter and distilled water, x = 21 cm, w_{SL} = 1.8 cm s^{-1} at different superficial gas velocities [261] (For symbols see Fig. 20)

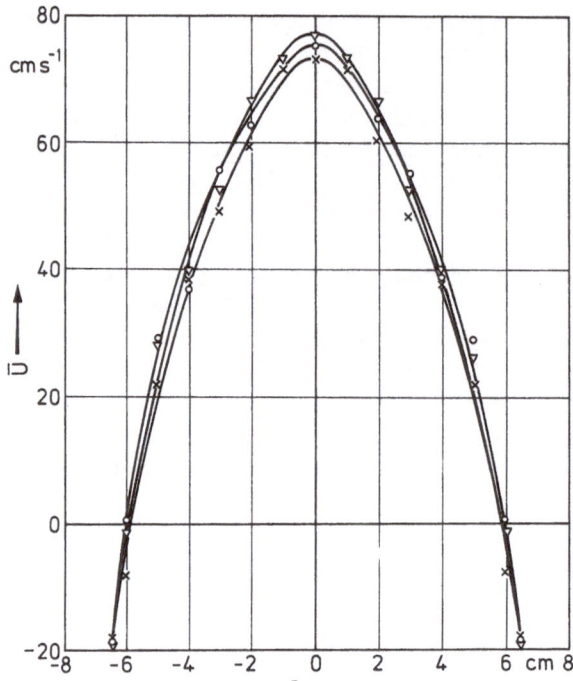

Fig. 22. True mean liquid velocity profiles, $\bar{U}$, in tower reactors using perforated plate 3.0 mm in hole diameter and distilled water, $x = 149$ cm, $w_{SL} = 1.8$ cm s^{-1} at different superficial gas velocities [261]. (For symbols see Fig. 20)

with much higher velocities than individual bubbles and the transition to a heterogeneous state occurs. When employing porous plates, $d_p \ll d_e$. Bubbles grow by coalescence, hence bubbles with broad diameter distribution are formed. Large bubbles are enriched in the column center and have much higher rise velocities than small ones. Since a close relationship exists between bubble velocity and mean liquid velocity [258], high local liquid velocities are attained in the column center. In systems with a perforated plate 1 mm in diameter, the difference between d_p and d_e is fairly small. Thus the coalescence effect is moderate; bubble size distribution is fairly narrow. Therefore the mean local liquid velocity profile is less dependent of gas velocity, than the one with a porous plate. Furthermore the velocity maximum in the column center is smaller than the one with a porous plate, but larger than the one with a perforated plate 3 mm in hole diameter. The profile of the local mean velocity, $\bar{U}$, is not influenced appreciably by the absence or presence of the bacterial cells, or by batch or continuous operation (Fig. 23). Also the radial profiles of turbulence intensity, u', and relative turbulence intensities, I, have the same shapes in the absence or presence of growing cells, in batch as well as in continuous operation (Fig. 24 and 25). The same holds true for the radial profiles of the relative mean gas hold-up, ε_G (Fig. 26), bubble diameters $\bar{d}$ and d_S (Fig. 27) and bubble rise velocities, u_B (Fig. 28) [261].

The radial profiles of macro time scales, τ_M, in these systems have parabolic shape with their minimum in the column center (Fig. 29). In cell free systems, this minimum is the broadest and has a much lower value (0.074 s) than those with growing cells (0.09 s in batch and 0.098 s in continuous operation) [102].

The radial profiles of micro time scales, τ_E, in these systems have similar shapes (Fig. 30). In the center of the column they are nearly constant. At the wall they steeply increase and at a distance of about 5 cm from the column center they pass a minimum. The dissipation time scale plateau in the column center is lower in the absence of cells $(0.382 \times 10^{-3}$ s) than in their presence $(0.4 \times 10^{-3}$ s in batch and 0.42×10^{-3} s in continuous operation).

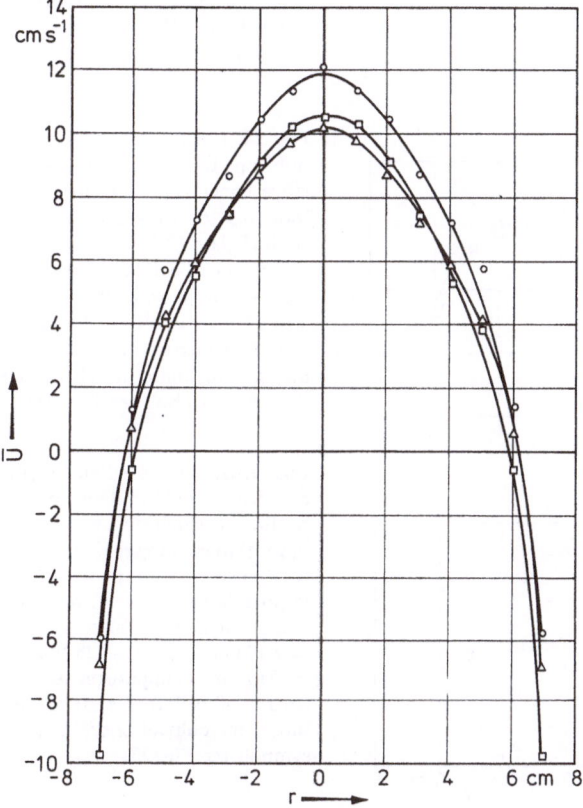

Fig. 23. True mean liquid velocity, $\bar{U}$, profiles in a single stage con-current (air lift) tower loop reactor (Reactor A) using porous plate 17.5 μm in pore diameter and *E. coli* cultivation medium (Table 8). $w_{SG} = 2.17$ cm s^{-1}, T = 28 °C, x = 170 cm. Comparison of properties of the cell free system with those of cultivations [261].
o   medium without cells
△   batch cultivation of *E. coli*
□   continuous cultivation of *E. coli*

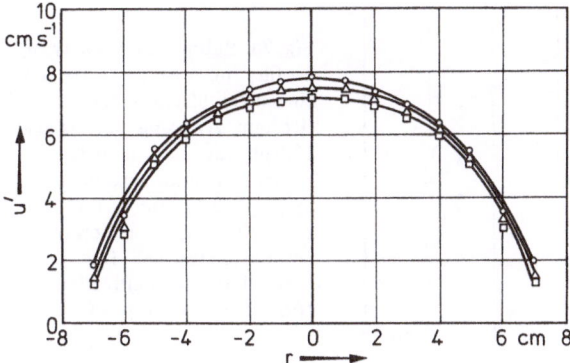

Fig. 24. Turbulence intensity, u', profiles in a single-stage concurrent (air lift) tower loop reactor (Reactor A) using porous plate 17.5 μm in pore diameter and *E. coli* cultivation medium (Table 8). w_{SG} = 2.17 cm s^{-1}, T = 28 °C, x = 170 cm.
Comparison of properties of the cell free system with those of cultivation [261]. (For symbols see Fig. 23)

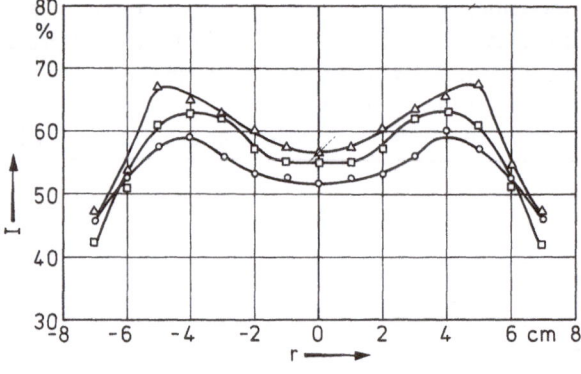

Fig. 25. Relative turbulence intensity, I, profiles in a single-stage concurrent (air lift) tower loop reactor (Reactor A) using porous plate 17.5 μm in pore diameter and *E. coli* cultivation medium (Table 8). $w_{SG} = 2.17$ cm s^{-1}, $T = 28$ °C, $x = 170$ cm. Comparison of properties of the cell free system with those of cultivations[261]. (For symbols see Fig. 23)

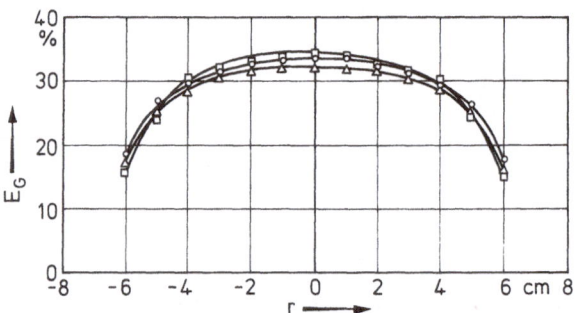

Fig. 26. Local mean relative gas hold up, E_G, profiles in a single stage concurrent (air lift) tower loop reactor (Reactor A) using porous plate 17.5 μm in pore diameter and *E. coli* cultivation medium (Table 8). $w_{SG} = 2.17$ cm s^{-1}, $T = 28$ °C, $x = 210$ cm. Comparison of properties of the cell free system with those of cultivations[261]. (For symbols see Fig. 23)

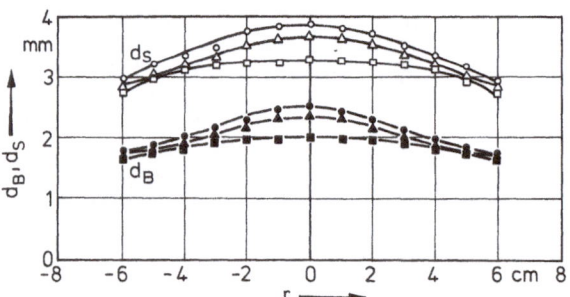

Fig. 27. Mean bubble diameter, d_B, and Sauter bubble diameter, d_S, profiles in a single stage concurrent (air lift) tower loop reactor (Reactor A) using porous plate 17.5 μm in pore diameter and *E. coli* cultivation medium (Table 8). $w_{SG} = 2.17$ cm s^{-1}, $T = 28$ °C, $x = 210$ cm. Comparison of properties of cell free system with those of cultivations[261]. (For symbols see Fig. 23)

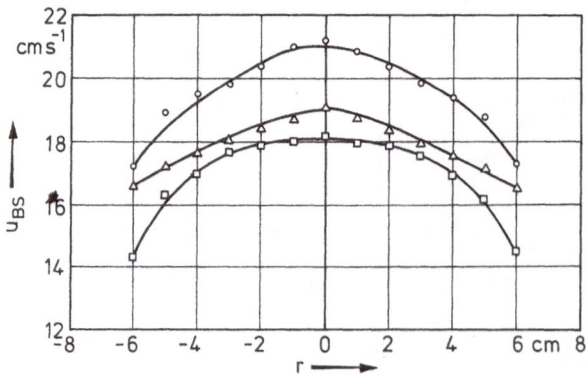

Fig. 28. Bubble rise velocity, u_{Bs}, profiles in a single stage concurrent (air lift) tower loop reactor (Reactor A) using porous plate 17.5 μm in pore diameter and *E. coli* cultivation medium (Table 8), $w_{SG} = 2.17$ cm s^{-1}, $T = 28$ °C, $x = 210$ cm. Comparison of properties of the cell free system with those of cultivations[261]. (For symbols see Fig. 23)

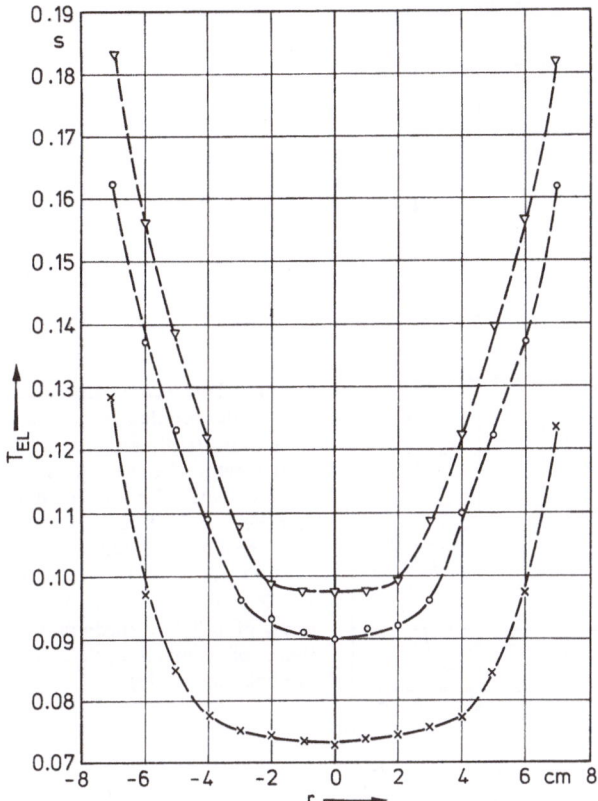

Fig. 29. Turbulence macro time scale, T_M, profiles in a single stage concurrent (air lift) tower loop reactor (Reactor A) using porous plate 17.5 µm in pore diameter and *E. coli* cultivation medium (Table 8). $w_{SG} = 2.17$ cm s^{-1}, $T = 28$ °C, $x = 80$ cm. Comparison of properties of the cell free system with those of cultivations [261].

× in absence of *E. coli*
○ batch cultivation of *E. coli*
▽ continuous cultivation of *E. coli*

According to the theory of Whalley and Davidson [290] and its extension by Joshi and Sharma [291], there are liquid circulation cells in bubble columns. The centrum of these cells corresponds to the radial position where $\bar{U} = 0$. One would expect at this position a minimum of τ_E and a maximum of the energy dissipation rate, E. Astonishingly, this minimum does not occur at this radial position, but at positive $\bar{U}$ values, where the $\bar{U}$-profile has a small irregularity (arrow in Fig. 31). It is not yet clear why this shift of the τ_E minimum occurs.

Anyhow, it is expected that in the range of this minimum τ_E, the gas dispersion is especially effective due to the maximum of E.

When employing perforated plates 1 or 3 mm in hole diameter, the power spectrum, $E_1(k)$ does not depend appreciably on w_{SG} [259, 260, 273] (Fig. 32). However, when using a porous plate, $E_1(k)$ is shifted to higher energy values and it attains a maximum with increasing w_{SG}. Then it diminishes, probably due to close bubble packing which damps the turbulence in the space between the bubbles (Fig. 33).

According to [1], the position and shape of the energy dissipation spectrum $\Phi_1(k)$ considerably influences the efficiency of the aerator with regard to the gas dispersion (OTR per unit power input: kg O_2/kWh). In Fig. 34, energy dissipation spectra, $\Phi_1(k)$'s are plotted as functions of the wave number, k, for different gas distributions. All of these spectra pass a maximum with increasing k. The higher k is at which this maximum occurs, and the narrower the spectrum, the

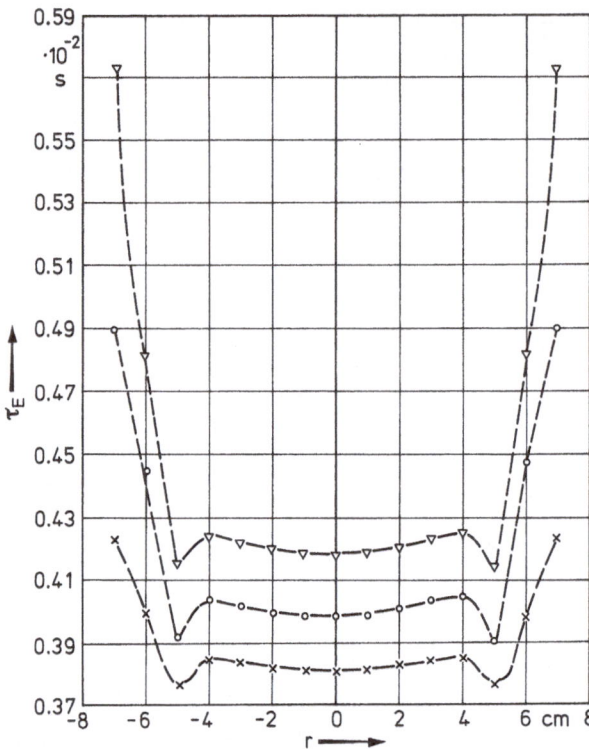

Fig. 30. Turbulence dissipation time scale profiles in a single stage concurrent (air lift) tower loop reactor (Reactor A) using porous plate 17.5 μm in pore diameter and *E. coli* cultivation medium (Table 8). w_{SG} = 2.17 cm s^{-1}, T = 28 °C, x = 80 cm. Comparison of properties of cell free medium with those of cultivations[261]. (For symbols see Fig. 29)

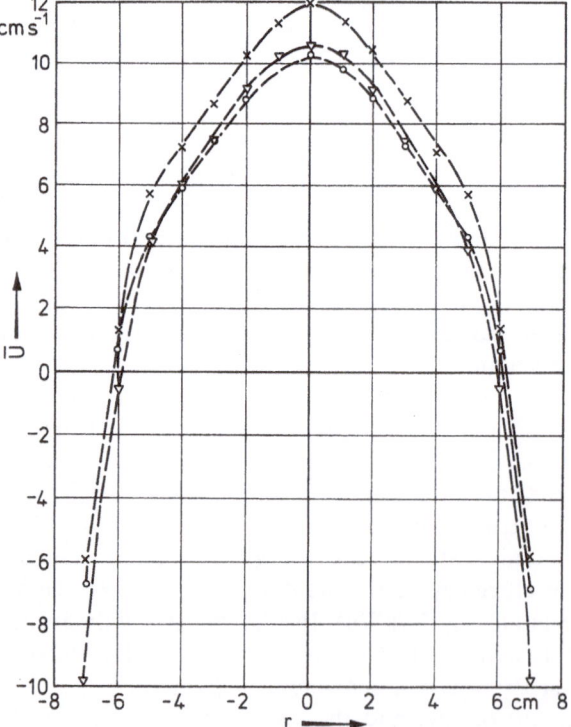

Fig. 31. True mean liquid velocity, Ū, profiles in a single stage concurrent (air lift) tower loop reactor (Reactor A) using porous plate 17.5 μm in pore diameter and *E. coli* cultivation medium (Table 8). w_{SG} = 2.17 cm s^{-1}, T = 28 °C, x = 30 cm. Comparison of properties of the cell free system with those of cultivations[261]. (For symbols see Fig. 29)

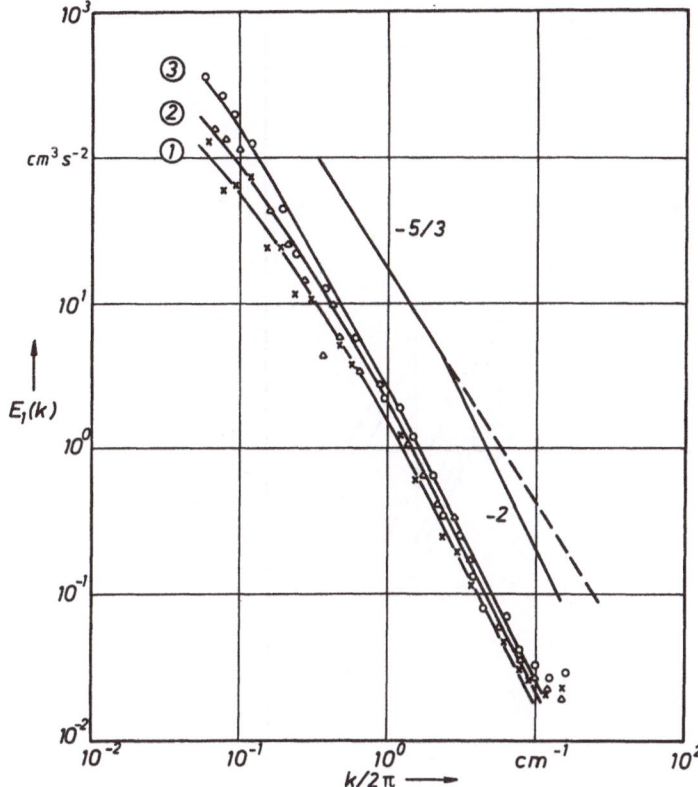

Fig. 32. One-dimensional power spectrum, $E_1(k)$, in a tower reactor using perforated plate 3 mm in diameter and 1% methanol solution [260, 273]; 1 $w_{SG} = 2.67$ cm s^{-1}; 2 $w_{SG} = 5.35$ cm s^{-1}; 3 $w_{SG} = 8.02$ cm s^{-1}

more efficient is the aerator. The explanation for this phenomenon is as follows: Since the gas dispersion can only occur at inverse wave numbers which are much smaller than the bubble diameter, i.e. $k^{-1} \ll d_B$, gas dispersion only occurs at very high k values. The higher the percentage of energy, which is dissipated in the range or the closer the maximum of $\Phi_1(k)$ is to this range, the higher is the efficiency of the aerator.

One can see from Fig. 34 that by using porous plate aerators, the maximum of $\Phi_1(k)$ is at the highest k value. When using a perforated plate 1 or 3 mm in hole diameter, the maximum position shifts to lower k values and the maxima are less significant. When increasing gas flow rate, $\Phi_1(k)$ is shifted to higher energy dissipation rates but its shape and maximum position is not influenced, i.e. its efficiency remains the same (see curves 1 and 4 in Fig. 34).

The one-dimensional power spectrum, $E_1(n)$, and the energy dissipation spectrum, $\Phi_1(n)$, change only slightly in a radial direction in the tower [261]. To describe these variations, different characteristics of these spectra were used. Sometimes the power spectrum can be described by a particular function in a wide range of n, e.g. $E_1(n) \sim n^{-2}$. In this case, the amplitude of the measured spectrum can be

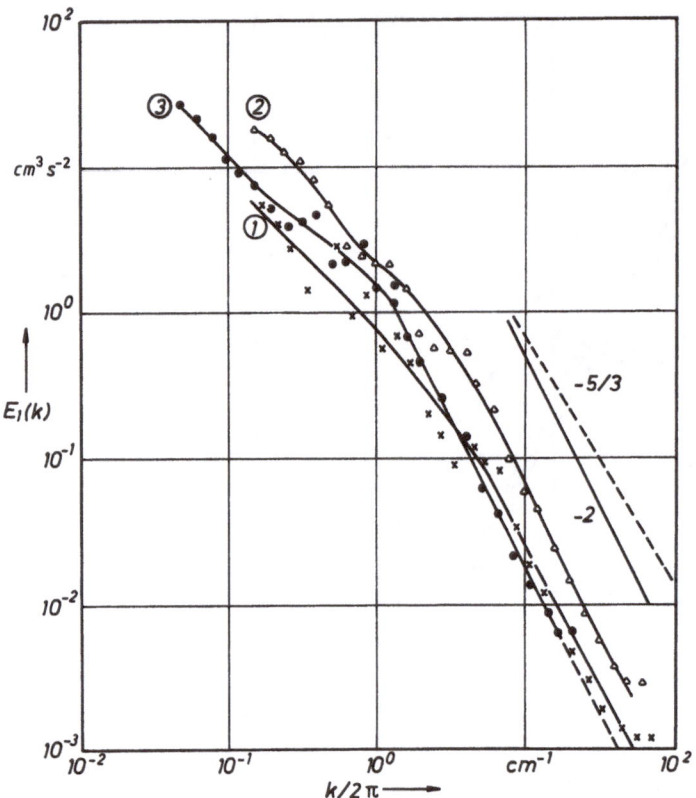

Fig. 33. One-dimensional power spectrum, $E_1(k)$, in a tower reactor using porous plate 5 μm in pore diameter and 1% methanol solution [260, 273]. (For symbols see Fig. 32)

characterized by its distance from a reference straight line (in a log-log plot) and the frequency range of the validity of this particular function. Such a description of $E_1(n)$ is sometimes possible in reactor B by employing *E. coli* medium (Table 8). In this case, $E_1(n) \sim n^{-2}$ for a wide range of n. Thus the radial variations of the frequency range, n_E, and amplitude, A_E, were considered. In the absence of cells in the tower center, n_E seems to have a maximum and A_E a minimum. In the presence of cells (batch cultivation of *E. coli*), n_E and A_E vary only slightly in the tower cross section. Unfortunately, $E_1(n) \sim n^{-b}$ generally holds true, where b is a function of n. With increasing n, b also increases. Thus this simple characterization cannot be employed.

The energy dissipation spectrum must have a maximum. If the measuring technique allows one to find this maximum, $\Phi_1(n)$ can be characterized by the frequency at which this maximum occurs, n_Φ, and by the height of the maximum, A_Φ. Again in reactor B, n_Φ and A_Φ could be determined by employing *E. coli* medium (Table 8). In the presence of cells (in batch as well as in continuous cultures), n_Φ and A_Φ are uniform in the tower cross sections. In the absence of cells, A_Φ is uniform and n_Φ changes slightly, in the tower center. In Reactor B, the measuring position was at a distance of 8 cm from the stage separating tray.

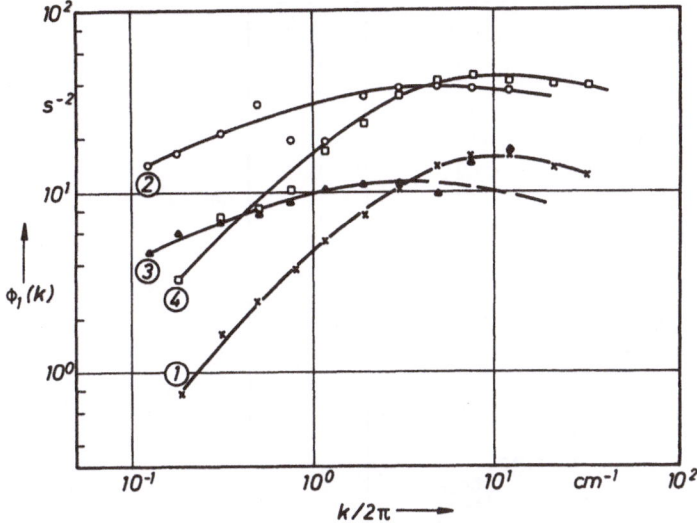

Fig. 34. Energy dissipation spectrum, $\Phi_1(k)$ in a tower reactor at $w_{SL} = 1.8$ cm s^{-1} and $r = 0$ (at the center) using 1% methanol, different aerators and superficial gas velocities [260, 273].
1 porous plate 5 μm
2 perforated plate 1 mm $w_{SG} = 2.67$ cm s^{-1}
3 perforated plate 3 mm
4 porous plate 5 μm $w_{SG} = 5.35$ cm s^{-1}

In Reactor A and at further distance ($x = 170$ cm) from the aerator, n_E and n_Φ seem to be fairly uniform; A_Φ seems to have a maximum in the center, A_E seems to oscillate.

At an intermediate distance ($x = 30$ cm) from the aerator, the position dependence of these parameters is not clear.

In Figs. 35 and 36, one-dimensional power spectra and energy dissipation spectra are shown which were measured at $x = 30$ cm and in the center ($r = 0$) of Reactor A using a porous plate 17.5 μm in pore diameter and *E. coli* cultivation medium (Table 8).

In Fig. 35, the dotted line represents $E_1(n) \sim n^{-2}$. One can recognize from Fig. 35 that between model medium and batch cultivation only a slight difference exists. Both of them have a course of about $E_1(n) \sim n^{-5/3}$ in the range $n = 8$ to 150 Hz and for $n > 150$ Hz they follow the course $E_1(n) \sim n^{-2}$. Fig. 36 again shows the slight differences between biological and model media for $n = 20$ to 500 Hz. One can recognize from this figure that the n-value at which $\Phi_1(n)$'s maximum prevails, n_Φ, is higher than 500 Hz, the upper limit which can be attained by the employed technique.

In Fig. 37 one-dimensional power spectra are plotted which were measured in Reactor B (distance from Tray: 8 cm, $r = 0$ cm) employing *E. coli* medium (Table 8). The dotted line again represents the function $E_1(n) \sim n^{-2}$. There are significant differences between the cellfree systems, batch and continuous cultivations of *E. coli*. In the relationship $E_1(n) \sim n^{-b}$, b continuously increases with increasing n from $b \simeq 0$ at $n \simeq 7$ Hz to about $b \simeq 3$ at $n = 500$ Hz. The large differences between bio-

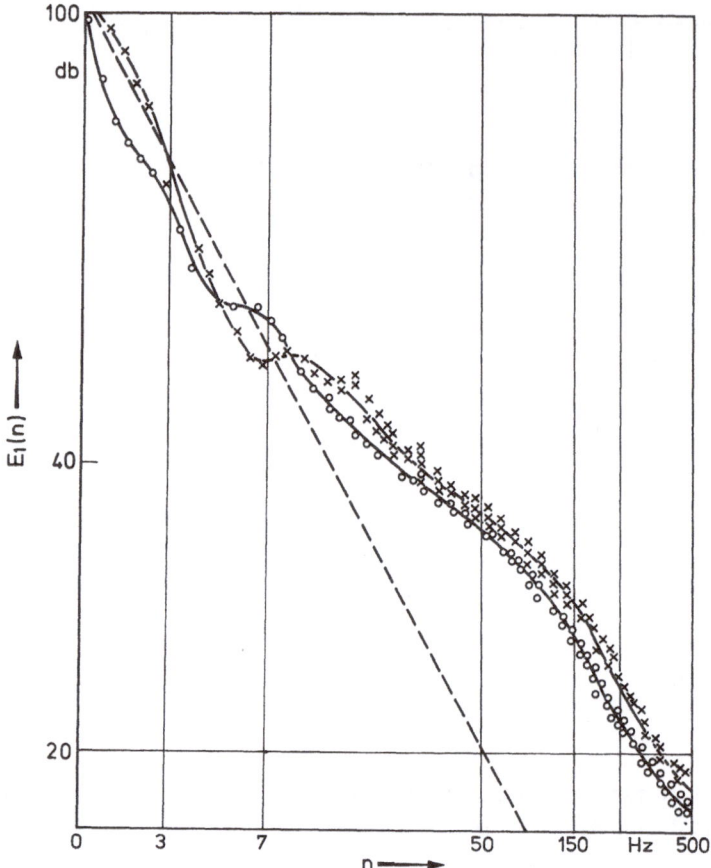

Fig. 35. One-dimensional power spectrum $E_1(n)$ in Reactor A using porous plate 17.5 μm in pore diameter and *E. coli* cultivation medium (Table 8). $w_{SG} = 2.17$ cm s^{-1}, T = 28 °C, x = 30 cm, r = 0 cm (tower center). Comparison of properties of the cell free system with those of batch culture [261]; ------: $E(n) \sim n^{-2}$. (For symbols see Fig. 29)

logical and model media can be recognized from Fig. 37b. In the absence of bacteria, $n_\Phi > 500$ Hz and thus it could not be determined. In a batch culture, n_Φ and A_Φ are higher than in a continuous culture. The drastic changes of $\Phi_1(n)$ in the presence of cells are probably due to the interaction of the foam with the stage separating trays. The low efficiency of Reactor B, when employing biological media, can be recognized from the low values of n_Φ and A_Φ in Fig. 37b (see also Chapter 7.2.2).

7 Reactor Performance

7.1 General Considerations

Bioreactors can be operated according to different strategies depending on the structure of the production costs. (Finishing costs are not considered here).

The following cases are considered:

A. Substrate conversion, U_s, should be maximized, if the raw material (substrate) costs dominate
B. Productivity, Pr, should be maximized if investment and running costs (with exception of substrate) are high
C. Oxygen conversion, U_O, should be maximized, if oxygen transfer costs are important
D. Specific power input should be minimized, if energy costs (for gas dispersion and cooling) are high.

7.1.1 *Case A* (Maximizing Substrate Conversion)

The lower the intensity of longitudinal medium dispersion, the higher the conversion in continuous bioreactors. However, at low dispersion intensity cell washout occurs.

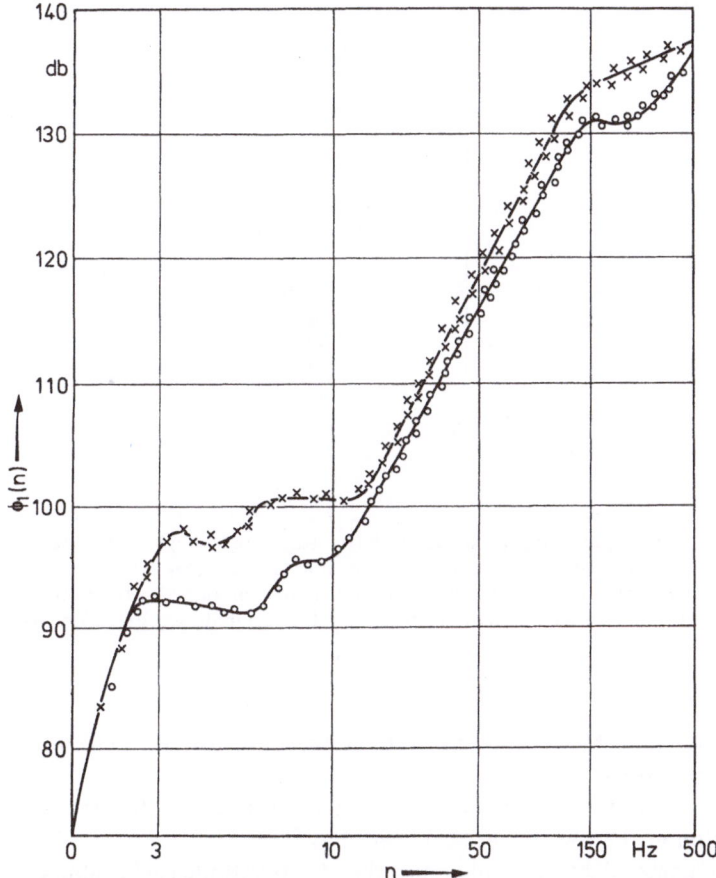

Fig. 36. Energy dissipation spectrum, $\Phi_1(n)$, in Reactor A using porous plate 17.5 µm in pore diameter and *E. coli* cultivation medium (Table 8). (For operation conditions see Fig. 35). $x = 30$ cm, $r = 0$ cm (tower center). Comparison of properties of the cell free system with those of batch culture [261]. (For symbols see Fig. 29)

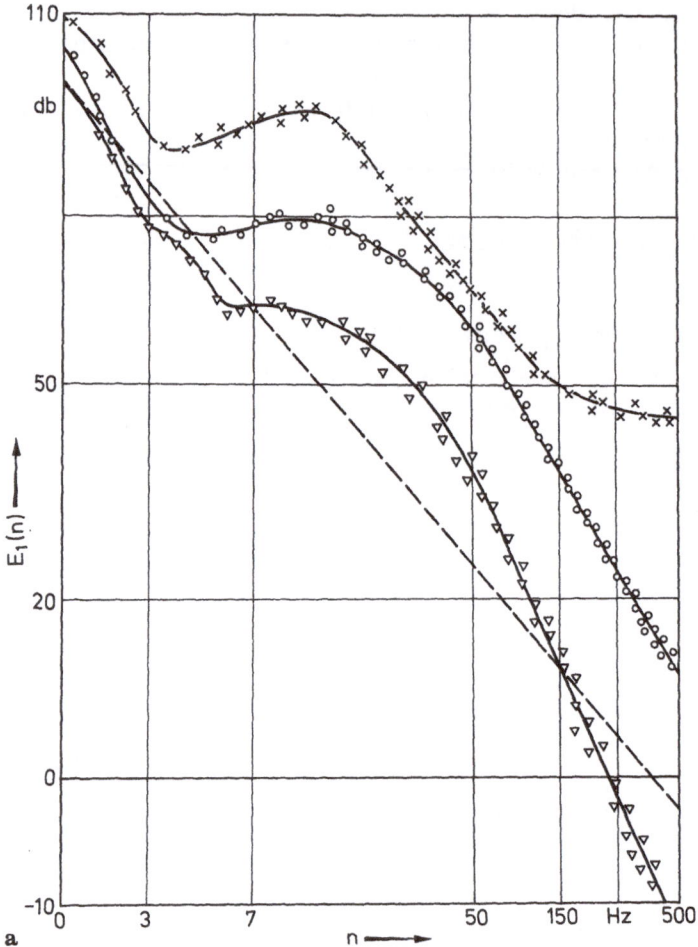

Fig. 37a. One-dimensional power spectrum $E_1(n)$ in reactor B (ten-stage tower loop) using perforated plate trays 3.0 mm in hole diameter and *E. coli* cultivation medium (Table 8), $w_{SG} = 2.17$ cm s^{-1}, $T = 28$ °C, distance from 5th tray: 8 cm, $r = 0$ (tower center). Comparison of properties of the cell free systems with those of batch and continuous culture[261]. (For symbols see Fig. 29); ------ : $E(n) \sim n^{-2}$

To avoid washout and to achieve high substrate conversion, tower reactors with medium recycling can be used.

It is possible to decouple the longitudinal dispersions of cells and substrate, if the substrate concentration at the tower exit, S_e, is very low[21].

Since the substrate passes the tower only once, the effective Bodenstein-number is given for it by $Bo_R = Bo(1 + \gamma)$. Bo_B increases with the medium recirculation ratio, γ. On the other hand, a large percentage of cells are recycled, thus the cell dispersion intensity increases with increasing γ. At high medium recirculation ratio ($\gamma > 20$), the tower loop behaviour approaches that of a continuous stirred tank

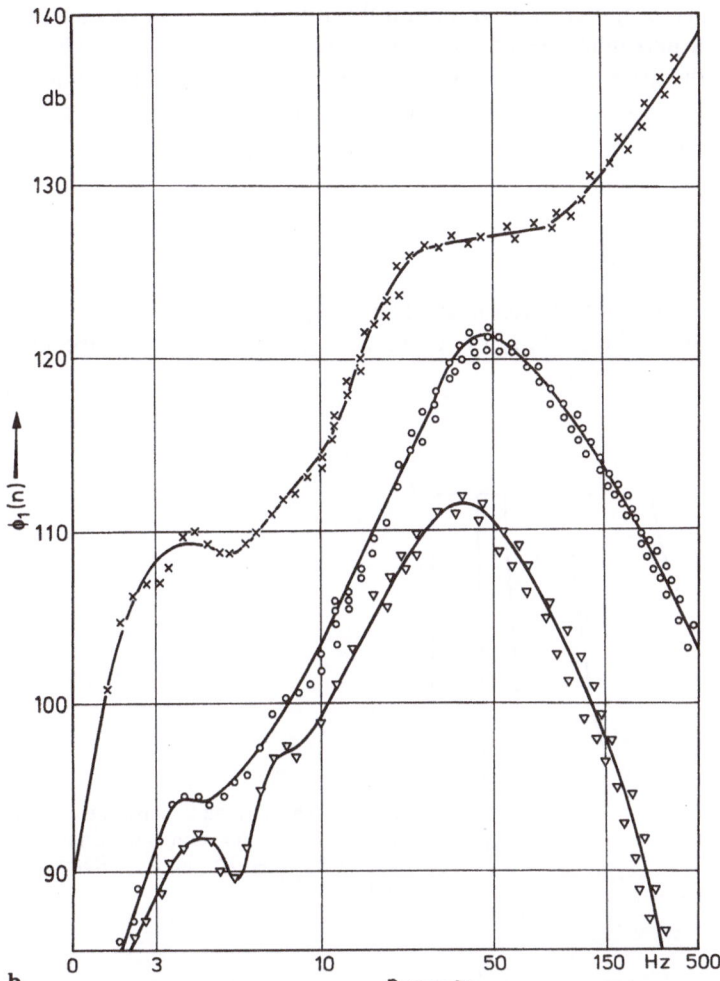

Fig. 37b. Energy dissipation spectrum, $\Phi_1(n)$, in reactor B (ten-stage tower loop) using perforated plate trays 3.0 mm in hole diameter and *E. coli* cultivation medium (Table 8). (For operation conditions see Fig. 37a). Distance from 5th tray: 8 cm, r = 0 (tower center). Comparison of properties of the cell free system with those of batch and continuous culture [261]. (For symbols see Fig. 29)

reactor, CSTR, with regard to the cells but the behaviour of a plug flow reactor, PFR, with regard to the substrate [21]. This decoupling holds true as long as $S_e \simeq 0$.

At too high γ values, $S_e > 0$, and the substrate is also recycled and the tower loop system also behaves as a CSTR with regard to the substrate. Thus, there is an optimum γ-value at which $C_s = \dfrac{S_e}{S_0}$ has a minimum and the conversion, $U_A = 1 - C_s$, has a maximum, as long as the Bo-number is high enough [21].

This problem was quantitatively considered by Adler [17, 295]. His treatment is based on the results of Chen [285, 296], Pasquali et al. [286] and Todt et al. [297]. If the recirculation ratio, γ, is zero, there is an optimum Bo-number, Bo_{opt}, at which C_s has a minimum and U_s a maximum. The position of this C_s-minimum (U_s-maximum) strongly depends on the dimensionless mean residence time, the Damköhler number, $Da = \mu_m \tau$, where τ is the mean medium residence time in the reactor (Fig. 38), as well as on the dimensionless substrate saturation constant, K (Fig. 39):

$$K = \frac{K_s}{S_o + X_o/Y_{X/S}} \qquad (102)$$

where X_0 is the cell mass concentration in the feed.

Below a critical Da-number, Da_{crit}, or above a critical K, K_{crit}, the optimum reactor is the CSTR.

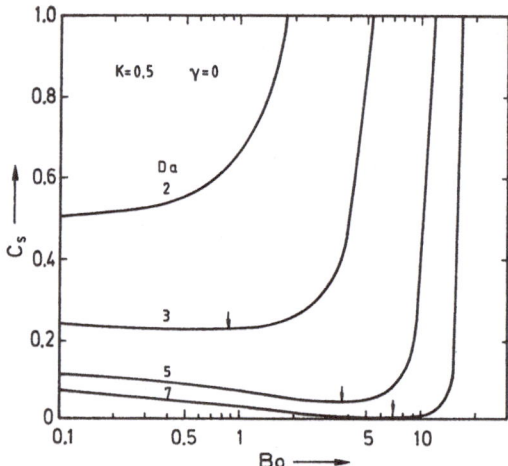

Fig. 38. Dimensionless substrate concentration, C_s, as a function of the Bo-number at $\gamma = 0$, $K = 0.5$ and for different Da-numbers [17, 295]

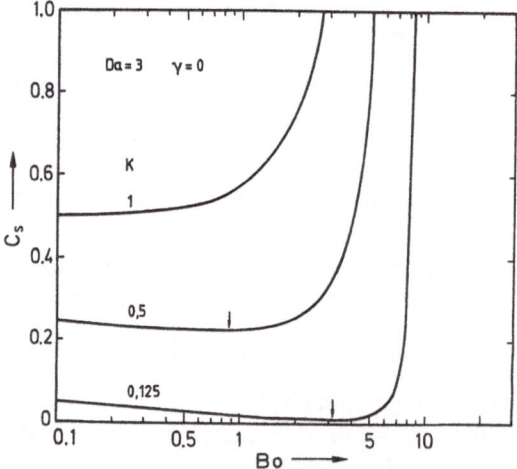

Fig. 39. Dimensionless substrate concentration, C_s, as a function of the Bo-number at $Da = 3$ and $\gamma = 0$ and for different saturation constants, K [17, 295]

For sterile feed ($X_0 = 0$) [17, 295]:

$$Da_{crit} = \cfrac{1}{1 - \cfrac{K}{\sqrt{K(K+1)}}} \tag{103}$$

and

$$K_{crit} = \cfrac{1}{\cfrac{1}{\left(1 - \cfrac{1}{Da}\right)^2} - 1} \tag{104}$$

Introduction of medium recirculation can improve C_s and/or U_s (Fig. 40), if $Bo > Bo_{opt}$, i.e. $\gamma_{opt} > 0$.

On the other hand for $Bo \leq Bo_{opt}$, $\gamma_{opt} = 0$.

Fig. 41 shows how the Bo-number influences γ_{opt}. Near Bo_{opt}, γ_{opt} considerably depends on the Bo-number. At large Bo-numbers, γ_{opt} approaches a constant value and C_s varies with Bo only slightly. Because of the non-linearity of the Monod model, γ_{opt} also exists slightly below Bo_{opt}. For comparison, the C_s values for CSTR and for $\gamma = 0$ with Bo_{opt} are also plotted in this figure. The transition from CSTR to dispersion model with Bo_{opt} and $\gamma = 0$ improves (reduces) C_s by 64%. The introduction of the medium recirculation at, e.g., $Bo = 15$, further improves (reduces) C_s by 33%. Altogether a 76% improvement of C_s can be achieved with regard to CSTR. From Fig. 38 one can recognize that for this case ($Bo = 15$) and for $\gamma = 0$ washout would occur. Washout can be avoided by recycling and C_s minimized and/or U_s maximized. In Fig. 42, an example is shown for C_s as a function of γ at $Bo = 7$ and $K = 1$ for different Da-numbers.

Since $Bo > Bo_{opt}$, γ_{opt}'s exist. For $Da \leq Da_{crit} = 3.41$, $\gamma_{opt} = \infty$ (CSTR). For $Da > D_{crit}$, γ_{opt}'s exist, at which C_s is considerably improved. Fig. 43 shows γ_{opt} as

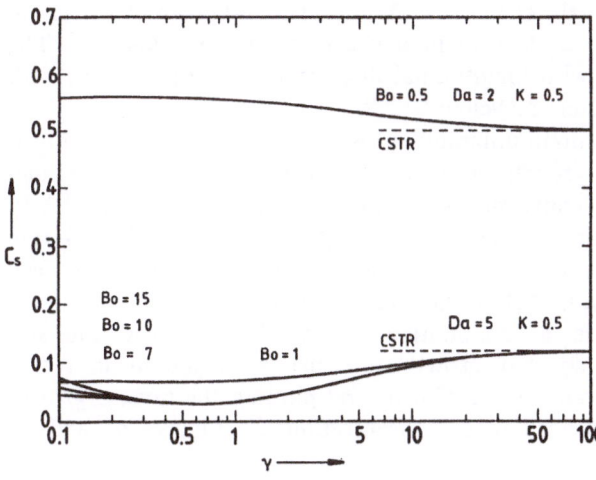

Fig. 40. Dimensionless substrate concentration, C_s, as a function of the medium recycling ratio, γ. Comparison of different reactor types [17, 295]

a function of the Da-number for two different K-values. For $Da > Da_{crit}$, γ_{opt} diminishes with increasing Da. In Fig. 44, γ_{opt} is plotted as a function of K for three different Da-numbers. For $K > K_{cri}$, γ_{opt} diminishes with increasing K.

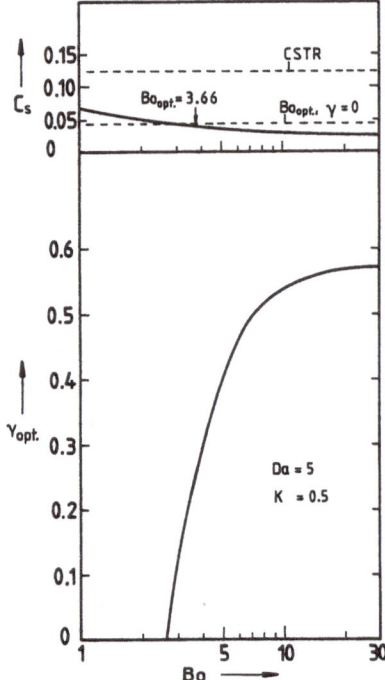

Fig. 41. Optimum recycling ratio, γ_{opt}, as a function of the Bo-number. Comparison of different reactor types [17, 295]

7.1.2 Case B (Maximizing cell mass productivity)

It is obvious that maximum cell productivity could be attained, if the cells could be cultivated in the nonlimited growth range, i.e. with μ_m. It is well-known that the non-limited growth in CSTR with sterile feed in a steady state is unstable [298]. The same is true of tower reactors with longitudinal dispersion [285, 286] as well as with longitudinal dispersion and medium recycling [17, 295].

However, it is possible to maintain unlimited growth at the entrance of the tower reactor and substrate limited growth at its exit, and by that to achieve high productivity and high substrate conversion in a stable steady state operation [21]. In this case the reactor behaviour can be described by the substrate limited rate equation. There is a formal analogy between autocatalytic reactions and cell growth with substrate limitation. Bishoff has shown that the maximum growth rate can be attained, if one uses a combination of a CSTR and PFR. The size of CSTR has to be chosen so that growth rate has a maximum in it [299]. The size of a CSTR can be evaluated graphically by plotting the inverse growth rate R_X^{*-1} as a function of the cell mass concentration, X. Figures 45 and 46 show such plots, where the dimensionless growth rate is

$$R_x = \frac{R_x^*}{X_o + Y_{X/S}S_o},$$ (105)

the growth rate according to Monod

$$R_x^* = \mu_m \frac{XS}{K_s + S}$$ (106)

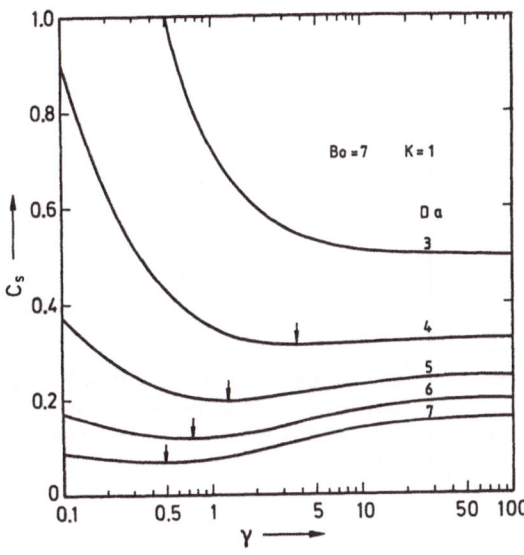

Fig. 42. Dimensionless substrate concentration, C_s, as a function of the recycling ratio, γ, at Bo = 7, K = 1 and for different Da-numbers [17, 295]

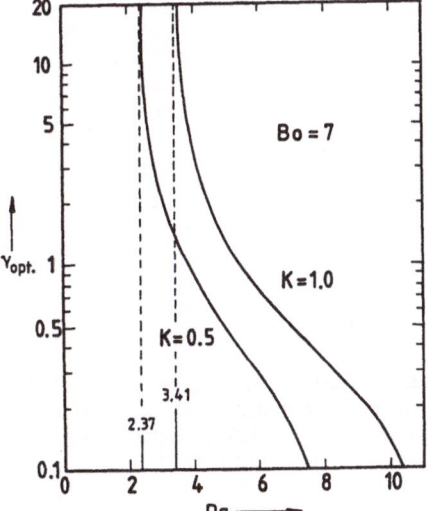

Fig. 43. Optimum recycling ratio, γ_{opt}, as a function of the Da-number at Bo = 7 for different saturation constants K [17, 295]

and the dimensionless cell mass concentration

$$C_x = \frac{X}{X_o + Y_{X/S}S_o} \cdot$$ (107)

One can recognize from Figs. 45 and 46 that R_x^{-1} passes a minimum at a particular C, which is called $C_{x\,crit}$.

The CSTR-PFR combination only exhibits an optimum, if $C_x > C_{x\,crit}$. For $C_x \leqq C_{x\,crit}$ the CSTR is the optimum reactor. $C_{x\,crit}$ is given by [17, 295]:

$$C_{x,\,crit} = (K + 1) - \sqrt{K(K + 1)} \geqq 0.5$$ (108)

i.e. for $C_x \leqq 0.5$, CSTR is always the optimum reactor.

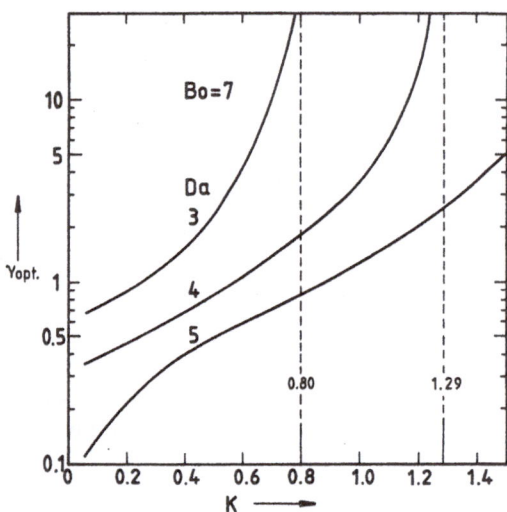

Fig. 44. Optimum recycling ratio, γ_{opt}, as a function of the saturation constant, K at Bo = 7 and for different Da-numbers [17, 295]

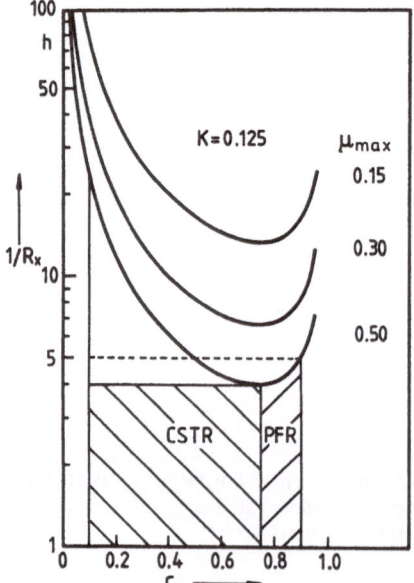

Fig. 45. Reciprocal related growth rate, R_x^{-1}, as a function of the dimensionless substrate concentration, C_x, at K = 0.125 for different μ_{max}. Combination of CSTR and PFR [17, 295]

If one uses a tower reactor with negligible longitudinal dispersion (PFR), the intensity of longitudinal dispersion can be controlled by employing medium recycling. There is an optimum recycling ratio, γ_{opt}, at which R_x has its maximum [17,295]. If one fixes the dimensionless cell concentration at the exit of the reactor, $C_{x,F}$, and looks for a reactor, which yields the smallest necessary volume, V_R, and mean residence time of the medium: $\tau = \dfrac{V_R}{V_L}$ or $Da = \tau\mu_m$, one finds that there is a PFR-loop combination with γ_{opt} at which Da has a minimum (Fig. 47), if $C_{xF} > C_{xF, crit}$, where

$$C_{xF, crit} = (K + 1) - \sqrt{K(K + 1)} \tag{109}$$

For $C_{xF} \leq C_{xF, crit}$, $\gamma_{opt} = \infty(CSTR)$

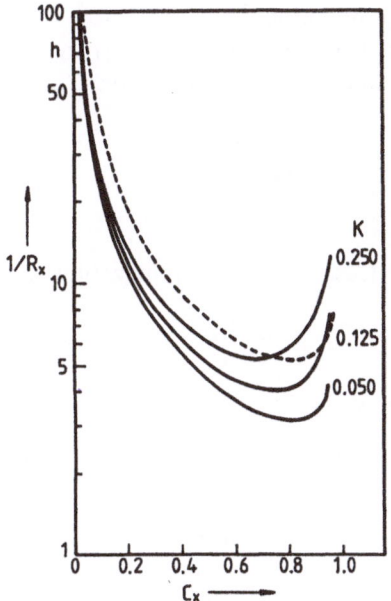

Fig. 46. Reciprocal related growth rate, R_x^{-1}, as a function of the dimensionless substrate concentration, C_x, at $\mu_{max} = 0.5 \, h^{-1}$ (———) and variable K, as well as at $\mu_m = 0.3 \, h^{-1}$ and $K = 0.05$ (------) [17,295]

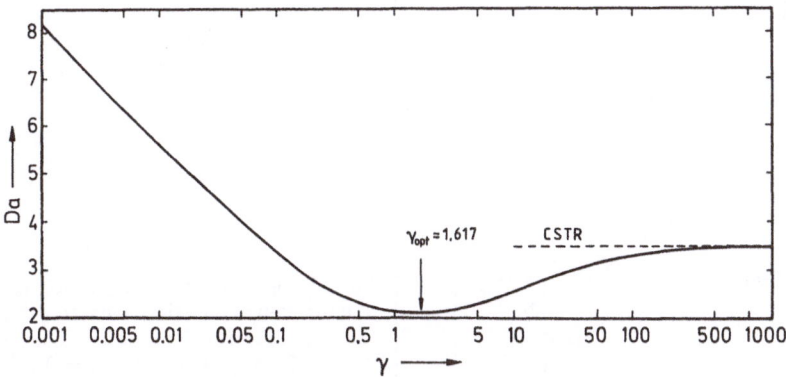

Fig. 47. Da-number as a function of the medium recycling ratio, for PFR with loop at $K = 0.125$, $C_{Xo} = 0$, $C_{XF} = 0.95$ [17,295]

For $C_{xF} > C_{x, F \, crit}$.

$$Da = (1 + \gamma)\left[\ln\frac{1+\gamma}{\eta+\gamma} + K \ln\frac{\zeta+\gamma}{\eta+\gamma}\right],\tag{110}$$

where

$$\eta = \frac{C_{xo}}{C_{xF}}$$

$$\zeta = \frac{1 - C_{xo}}{1 - C_{xF}}$$

γ_{opt} considerably depends on the exit cell mass concentration, C_{xF} (Fig. 48). With decreasing C_{xF}, γ_{opt} increases, and for $C_{xF} \leq C_{xF \, crit}$, $\gamma_{opt} = \infty$.

γ_{opt} also depends on K: with increasing K, γ_{opt} diminishes [17].

When employing a tower reactor with longitudinal dispersion and medium recycling, the relationships of case A can be used, since the exit cell mass concentration, C_{xF}, can also be written as

$$C_{xF} = \frac{X_F}{X_o + Y_{X/S}S_o} = \frac{Y_{X/S}(S_o - S_e) + X_o}{X_o + Y_{X/S}S_o} = \frac{S_o - S_e}{S_o} = U_s\tag{111}$$

Thus, reactor optimization with regard to U_s (Case A) is also optimization with regard to C_{xF}.

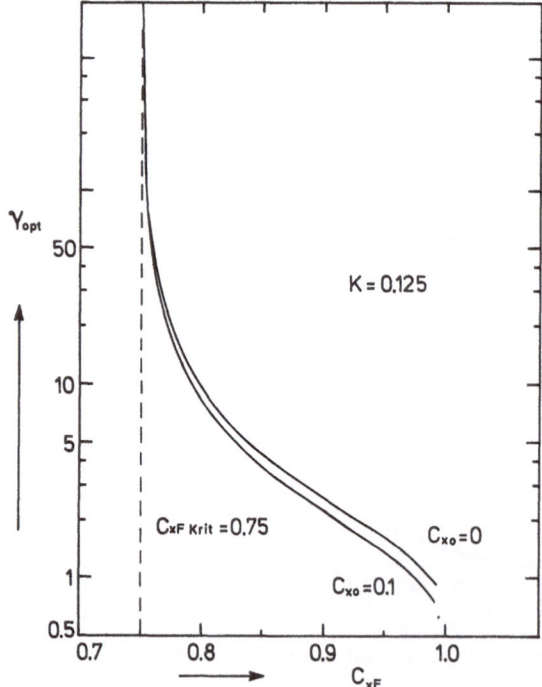

Fig. 48. Optimum recycling ratio, γ_{opt}, as a function of the exit cell mass concentration, C_{XF}, for PFR with loop at $K = 0.125$, $C_{Xo} = 0$ and $C_{Xo} = 0.1$ [17, 295)]

7.1.3 Case C (Maximizing Oxygen Conversion)

To maximize the oxygen conversion, U_O,

$$U_O = \frac{DO_{Lo} + OTR - DO_{Le}}{DO_{Lo} + OTR} = \frac{OUR}{DO_{Lo} + OTR}, \tag{112}$$

where O_{Lo} is the dissolved oxygen concentration in the medium at the medium entrance

O_{Le} the dissolved oxygen concentration in the medium at the medium exit.

$$OTR = \int_{z=0}^{z=1} k_L a \, (O_L^* - O_L) \, dz$$

$$OUR = \int_{z=0}^{z=1} \frac{1}{Y_{X/O}} \, \mu X \, dz,$$

the growth must be oxygen transfer limited, at least at the medium exit. For a CSTR, Eq. (112) reduces to

$$U_O = \frac{DO_{Lo} + k_L a \, (O_{Le}^* - O_{Le}) - DO_{Le}}{DO_{Lo} + k_L a \, (O_{Le}^* - O_{Le})} = \frac{\mu X / Y_{X/O}}{DO_{Lo} + k_L a \, (O_{Le}^* - O_{Le})} \tag{113}$$

For a given O_{Lo} and OTR, the oxygen conversion approaches unity if $O_{Le} \to 0$.

This leads to a similar problem which was considered in Case A. The most efficient use of dissolved oxygen would be in a PFR. However, because of the cell washout, medium recycling is necessary. In general, a type A tower loop reactor with longitudinal mixing and medium recycling can be used.

However, the theoretical treatment of this system is more complex than the one considered in Case A, since oxygen is transfered into the medium along the column. Particular systems (batch and fed batch tower loop reactors in oxygen transfer limited growth range) were treated in Refs. [181, 269-271]. In Chapter 5.2, the batch growth is considered.

By means of medium recycling it is possible to decouple the longitudinal dispersion of cell mass and substrate concentrations from the dissolved oxygen concentration. If the dissolved oxygen concentration, O_L, reduces to zero in the loop, then at high recirculation ratios, γ, the CSTR prevails with regard to the cell mass and substrate concentration and the PFR with regard to dissolved oxygen concentration [21]. Again to too high γ values, the dissolved oxygen is also recycled and its conversion is reduced. One can assume, that similar to Case A, Bo_{opt} as well as γ_{opt} exist, at which O_{Le} has a minimum, and U_O has a maximum. However, this problem has not been treated quantitatively yet.

7.1.4 Case D (Minimizing Specific Power Input)

In the oxygen transfer limited growth range the cell productivity is controlled by the OTR. In a CSTR, for the maximum productivity, Pr_m, Eq. (114) holds true:

$$Pr_m \simeq Y_{X/O} k_L a (O_L^* - O_L^c) \tag{114}$$

since usually

$$D(O_{Lo} - O_L^c) \ll k_L a (O_L^* - O_L^c) \tag{115}$$

Because of the low oxygen solubility, the maximum driving force $(O_L^* - O_L^c)$ is low. $Y_{X/O}$ is determined by the microorganism and the substrate. Thus Pr_m can usually be controlled by $k_L a$. Since the variation range of k_L is narrow, the specific gas/liquid interfacial area, a, is the main controlling parameter.

According to Eq. (35), a depends on E_G and d_s. Since E_G is also a function of d_s, the Sauter diameter is the primary variable. d_s depends in tower reactors on the primary bubble diameter, d_p, and on the coalescence rate, R_c, for which a simple linear relationship (106) is assumed:

$$R_c = \frac{dd_s}{dt} \simeq k_c (d_e - d_s) E_G \tag{116}$$

In Eq. (116) E_G is a function of d_s,

$\quad\quad\quad\quad k_c$ is the coalescence rate constant, which is strongly influenced by the medium character,

$\quad\quad\quad\quad d_e$ the dynamic equilibrium bubble diameter, which is constant in a single stage tower, except the aerator range, and

$\quad\quad(d_e - d_s)$ the driving force for the coalescence, which approaches zero with increasing time.

The secondary conditions are given by

$$d_s = d_p \qquad \text{for } t = 0 \text{, (at the aerator, } x = 0) \tag{116a}$$

$$d_s = d_e \qquad \text{for } t \to \infty \text{ (in large distance from the aerator } x \to \infty) \tag{116b}$$

The primary bubble diameter, d_p, is controlled by the primary gas dispersion at the aerator. Since the gas dispersion is caused by turbulence forces also in tower loop reactors [280], a close connection between the local turbulence energy dissipation rate and the local gas dispersion rate is expected. However, only relationships between the integral values of the energy dissipation rate, the integral specific power input, P_L/V_L, and the mean value of dynamic equilibrium bubble diameter, d_e, are known (e.g. 301, for a physical explanation see Chapter 4.3)

$$d_e = C_1 \frac{\sigma^{0.4}}{\varrho_L^{0.2}} \left(\frac{P_L}{V_L}\right)^{-0.4} \tag{117}$$

where C_1 is a constant.

One can also employ Eq. (117) for the primary gas dispersion

$$d_p = C_2 \frac{\sigma^{0.4}}{\varrho_L^{0.2}} \left(\frac{E}{V_d}\right)^{-0.4}, \tag{118}$$

where E is the local energy dissipation rate in the volume of primary gas dispersion,

V_d the volume of the primary gas dispersion and

C_2 a constant.

Equation (118) does not consider the differences in efficiencies of the aerator. It assumes that the fraction of microeddies is the same in the turbulence flow and it does not depend on the aerator. However, there is strong evidence that aerators have very different efficiencies [302] and these differences are due to the different energy dissipation spectra of turbulence, which are produced by these aerators [1]. To minimize the specific power input, the aerator efficiency has to be maximized. This holds true for primary dispersion. To minimize the specific power input, P_L/V_L, the coalescence rate, R_c, has to be minimized as well. Different strategies for minimizing R_c were considered in Ref. [303].

7.2 Particular Tower Loop Reactors

In Chapter 7.2, particular reactors will be considered separately and in Chapter 7.3, compared.

7.2.1 Single-stage Concurrent Tower Loop Reactor (Reactor A)

Two microorganisms were cultivated: *Hansenula polymorpha* and *Escherichia coli*; the yeasts in batch, fed batch and extended culture in the absence and presence of antifoam agents (Desmophen, soy oil), and the bacteria in batch and continuous culture in the presence of antifoam agent (Desmophen).

The *H. polymorpha* cultivations were computer simulated by means of a dispersion model (see 5.2) considering the space dependence of $k_L a$ according to Eq. (97) and calculating the longitudinal liquid dispersion coefficients, D_L, and gas dispersion coefficients, D_G, by means of Eq (119), which was recommended by Badura et al. [293]

$$D_F(t) = D_F^N(\tilde{w}_{SG}^E(t))^{0.33} \tag{119}$$

and by Eq. (120), which was recommended by Mangarz et al. [294].

$$D_G(t) = D_G^N(\tilde{w}_G^E(t))^{3.0} \tag{120}$$

In Eqs. (119) and (120):

$$D_L^N = 2.4 \times 10^{-4} \, m^2 \, s^{-1} \, (d_R \times 10^2 \, m^{-1})^{1.4}$$

and

$$D_G^N = 5.0 \times 10^{-8} \, m^2 \, s^{-1} \, (d_R \times 10^2 \, m^{-1})^{1.5}$$

where $\tilde{w}_{SG}^E(t)$ is the dimensionless superficial gas velocity at the gas entrance with regard to $w_{SG} = 0.01 \text{ m s}^{-1}$,

$$\tilde{w}_G^E(t) = \frac{\tilde{w}_{SG}^E(t)}{E_G}$$ the dimensionless effective gas velocity at the gas entrance, and

d_R the tower diameter $= 0.15 \text{ m}$

In Fig. 49, some measured and calculated dissolved oxygen longitudinal concentration profiles are shown. These profiles were evaluated by the dispersion model (Chapter 5.2) and by Eqs. (97), (119) and (120) for extended culture of *H. polymorpha* at a constant ethanol substrate concentration ($S = 5 \text{ g l}^{-1}$). By fitting the calculated profiles to the measured ones $k_L a^E$, $k_L a^\alpha$ and K_{ST} in Eq. (97) were identified. The specific surface area, a, was calculated by Eq. (35), by means of measured E_G and d_s values (see Chapters 4.3.1 to 4.3.4). The mass transfer coefficient, k_L, was evaluated by Eq. (117):

$$k_L = \frac{k_L a^\alpha}{a} \tag{121}$$

According to Eq. (121) it was assumed that the geometrical surface area, a, calculated by Eq. (35) and the mass transfer active area a, in $k_L a$ are identical.

Non-limited and oxygen transfer limited growth

The measurements employing *H. polymorpha* [16, 153, 154, 155, 181, 282] indicated that at the employed medium recycling rates ($\dot{V}_R = 1000$ to 2000 l h^{-1}) the tower

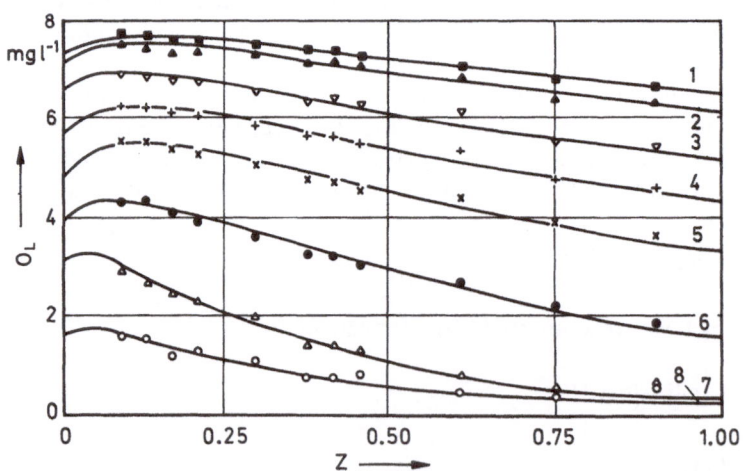

Fig. 49. Longitudinal concentration profiles of dissolved oxygen, DOC (mg l^{-1}) in Reactor A during *H. polymorpha cultivation at different cultivation times*, t. Substrate concentration: $S = 5 \text{ g l}^{-1}$ ethanol [16, 154].

1 ■ t = 2.0 h	5 × t = 13.0 h
2 ▲ t = 6.0 h	6 ● t = 14.0 h
3 ▽ t = 10.0 h	7 ▵ t = 17.0 h
4 + t = 12.0 h	8 ○ t = 25.0 h

reactor exhibited CSTR behaviour with regard to the cell mass, X, and substrate, S, concentrations. The longitudinal concentration profiles of dissolved oxygen are non-uniform and can be described by a dispersion model with particular Bo-numbers and space dependent $k_L a$.

The profiles were fairly uniform in nonlimited growth due to the low oxygen uptake rates (curves 1 to 5 in Fig. 49), and in the strongly oxygen transfer limited range due to the high oxygen uptake rate (curve 8). At the beginning of oxygen transfer limitation (curve 7 in Fig. 49 and curve 1 to 3 in Fig. 50), the nonuniformity of the profiles is most significant.

The dissolved oxygen was not recycled in the oxygen transfer limited growth range, because it was consumed in the loop as can be seen in Fig. 51 (upper part), in which the dissolved oxygen saturation is plotted as a function of the cultivation time, t, measured for *Hansenula polymorpha* on ethanol substrate at three different positions. At the end of the loop (z = −0), no oxygen could be detected even at t = 14 h, that is before the oxygen transfer limitation begins (t = 15 h, Fig. 51 lower part).

Also when using *E. coli* on glucose substrate [17, 305], the O_2-profiles were flat in nonlimited growth (curves 1 and 2 in Fig. 52). The dissolved oxygen concentration at the end of the loop (z = 0) already approached zero at t = 3.5 h (curve 3 in Fig. 52).

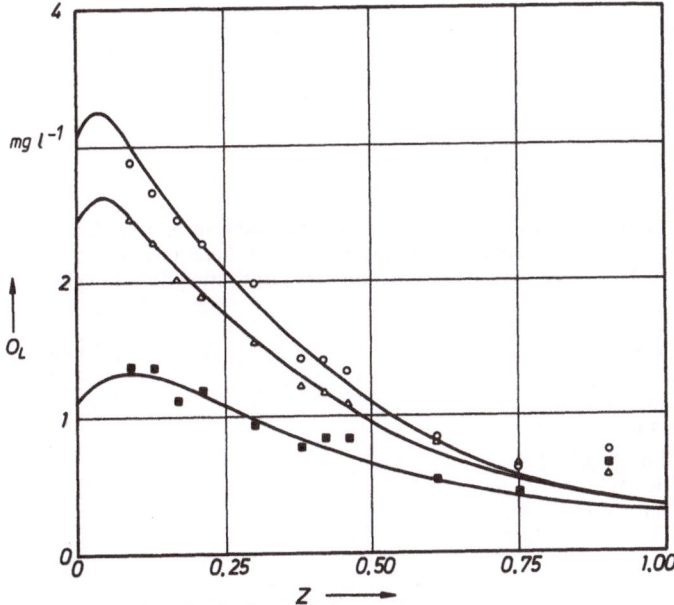

Fig. 50. Longitudinal concentration profiles of dissolved oxygen, DOC (mg l⁻¹) in Reactor A during *H. polymorpha* cultivation at different cultivation times, t. Substrate concentration: 5 g l⁻¹ ethanol [16, 154].

1 o t = 17.0 h	}	under strong oxygen
2 ▵ t = 19.0 h	}	transfer limited
3 ▪ t = 28.0 h	}	growth

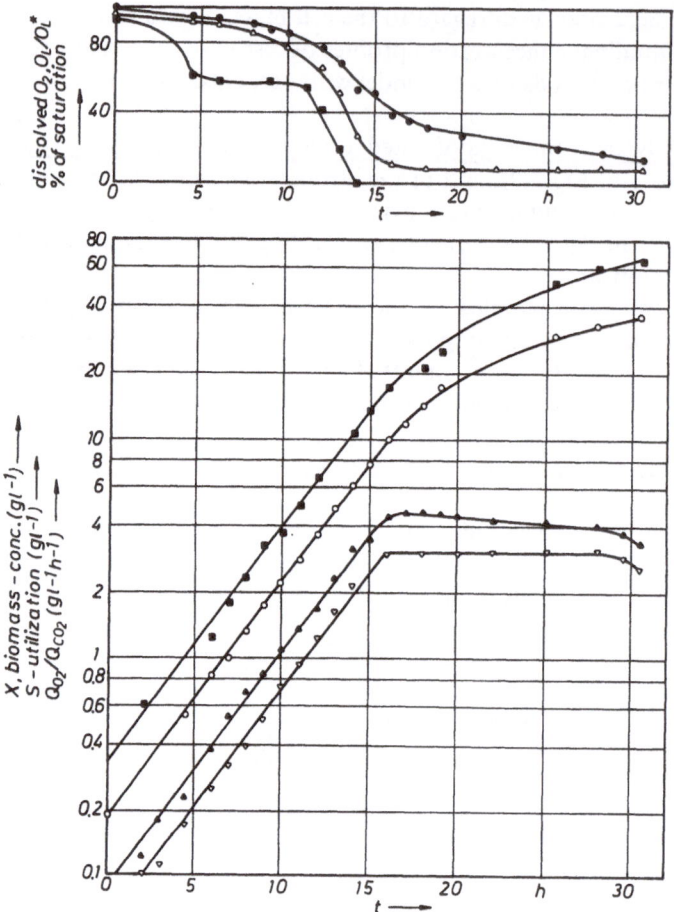

Fig. 51. *H. polymorpha* cultivation in Reactor A employing substrate ethanol in extended culture operation. Substrate concentration: 5 g l^{-1} kept constant by substrate feed. Aeration rate 0.55 vvm. *Upper part:* relative saturation of dissolved oxygen as a function of the cultivation time. Longitudinal position of the O_2-probes:

- ■ $z = -0$ (just below the aerator)
- ● $z = 0.09$ (at the aerator)
- △ $z = 0.90$ (at the tower head)

lower part: variation of the cultivation with time

- ■ substrate uptake rate (g l^{-1} h^{-1})
- ○ (dry) cell mass concentration, X (g l^{-1})
- ▲ oxygen uptake rate (g l^{-1} h^{-1})
- ▽ CO_2 production rate (g l^{-1} h^{-1})

By using the dispersion model as well as Eqs. (97) and (119), the dissolved oxygen saturation profiles were calculated and fitted to the measured ones, thus $k_L a^E$, $k_L a^\alpha$ and K_{ST} were identified. In Fig. 53, $k_L a^\alpha$ is shown as a function of the cultivation time for *H. polymorpha* and ethanol substrate [16, 153, 154, 181]. After inoculation, $k_L a^\alpha$ drops to low values, then quickly increases, at t ≃ 8 h, passes a maximum and diminishes at first fairly rapidly and after 14 h, gradually. Because the variations of the local d_S and E_G-values were slight in the tower (except for the aerator range, in which k_La^E drops to k_La^α), it was possible to characterize the two phase flow in the tower with local measurements of d_S and E_G. In Fig. 54, the specific interfacial area, a, calculated by Eq. (35), is plotted as a function of t for same run for which k_La^α is shown in Fig. 53. An initial drop, then a quick increase is common for k_La^α and a. However, k_La^α passes a maximum earlier and drops rapidly, while a attains its maximum later and diminishes only slightly.

The mass transfer coefficient, k_L, was calculated by Eq. (121) (Fig. 55). For $t \le 4$ h it is constant, but soon after it gradually diminishes for $t \le 15$ h, somewhat quicker, then-slighter [16, 153, 154, 181]. It is interesting to compare these properties of

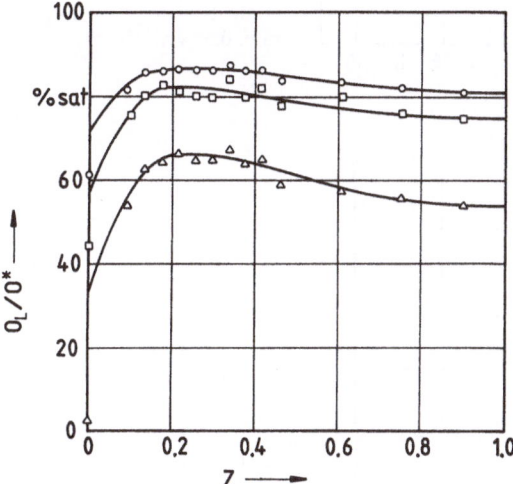

Fig. 52. Longitudinal concentration profiles of dissolved oxygen with regard to the saturation (%) in Reactor A during *E. coli* cultivation in batch operation. $\dot{V}_R = 1100\,l\,h^{-1}$, at different cultivation times [17, 315].

o   t =   30 min
△   t = 115 min
□   t = 210 min

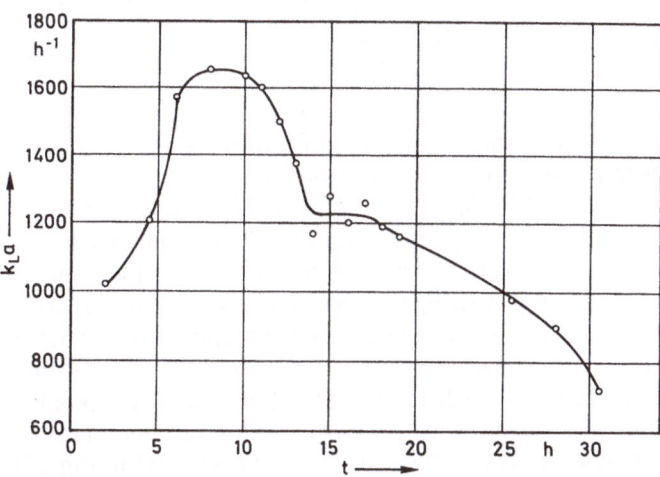

Fig. 53. Volumetric mass transfer coefficient, k_La, as a function of the cultivation time in Reactor A during the cultivation of *H. polymorpha* on ethanol substrate, $S = 5\,g\,l^{-1}$. Aeration rate: 0.55 vvm [16, 154]

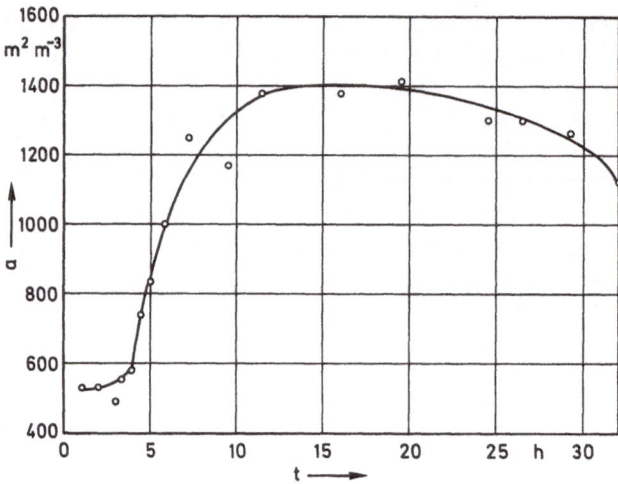

Fig. 54. Specific interfacial area, a, as a function of the cultivation time, t, in Reactor A during the cultivation of *H. polymorpha* on ethanol substrate, $S = 5$ g 1^{-1}. Aeration rate: 0.55 vvm [16, 154]

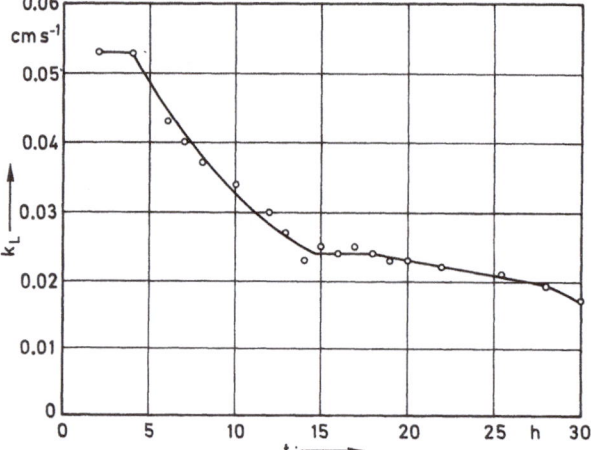

Fig. 55. Mass transfer coefficient, k_L, as a function of the cultivation time, t, in Reactor A during the cultivation of *H. polymorpha* on ethanol substrate, $S = 5$ g 1^{-1}. Aeration rate: 0.55 vvm [16, 154]

an antifoam free medium (*H. polymorpha*) with those of a medium (*E. coli*) in the presence of an antifoam agent (Desmophen). (The $k_L a$ values were calculated for the latter by means of O_2 balance) [17, 305]. In Fig. 56, $k_L a$ and k_L are plotted as a function of t. Both diminish with increasing cultivation time. Their courses are very similar because of d_S and E_G, thus a varies only slightly with t. Also their spacial variations in the tower are slight (Fig. 57). In the antifoam free system $k_L a$ drops from ca. 1600 to 800 h^{-1} (Fig. 53), and in presence of Desmophen from 2100 to 650 h^{-1} (Fig. 56). In both of these systems a varies from ca. 1400 to 1200 m^{-1} (Fig. 54 and Fig. 57). However, k_L is much lower in the presence of Desmophen ($k_L \simeq 6 \times 10^{-3}$ cm s^{-1} for $t > 60$ min; Fig. 56) than in its absence ($k_L \simeq 50 \times 10^{-3}$ cm s^{-1} for $t < 4$ h and $k_L \simeq 20 \times 10^{-3}$ cm s^{-1} for $t > 15$ h; Fig. 55). At low aeration rate, at which coalescence does not yet dominate, one can determine the ratio of k_L values from the corresponding ratio of $k_L a$ values. Such measurements were carried out after the cultivation was stopped in a small twin bubble

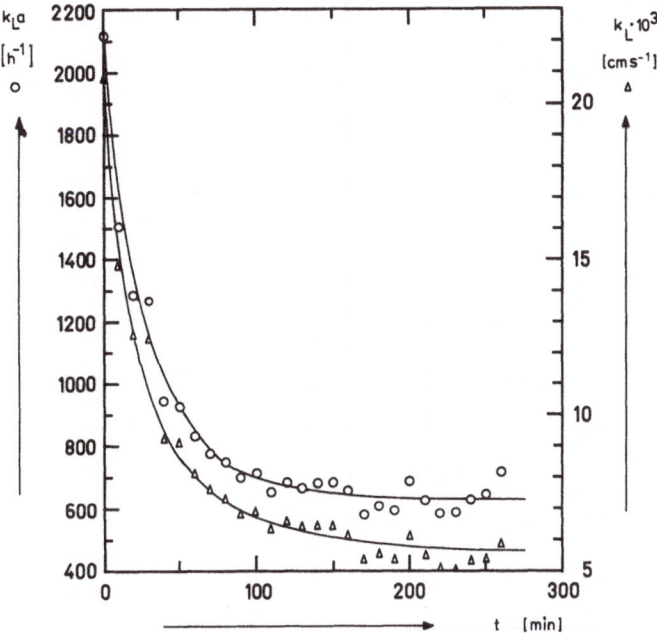

Fig. 56. Mass transfer coefficient, k_L, and volumetric mass transfer coefficient, k_La, as a function of the cultivation time in Reactor A during the cultivation of *E. coli* in batch operation. $\dot{V}_R$ = 1051 l h^{-1} [17, 305]

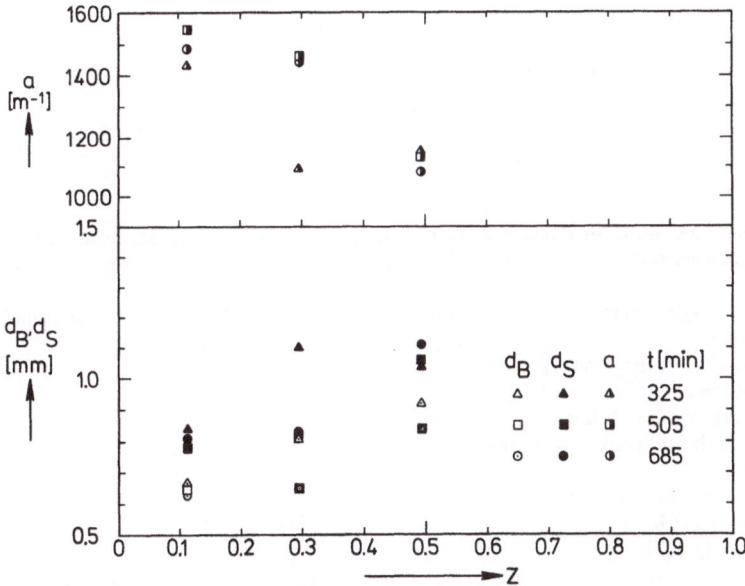

Fig. 57. Specific interfacial area, a, mean bubble diameter, d_B, and Sauter bubble diameter, d_S, as a function of the longitudinal position in Reactor A during the cultivation of *E. coli* at different times in continuous culture at high recycling ratio [17, 305]

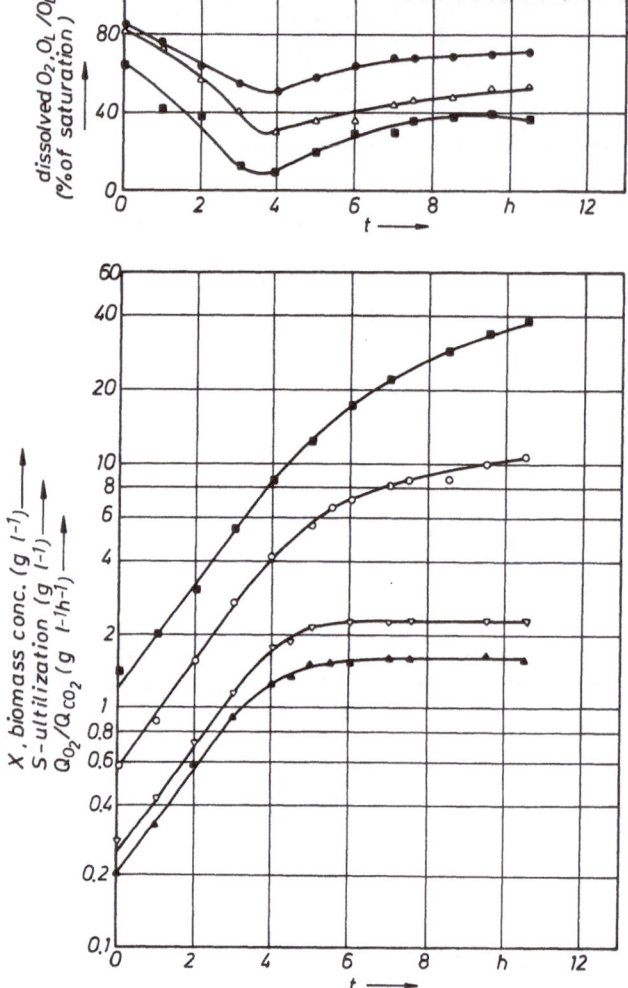

Fig. 58. *H. polymorpha* cultivation in Reactor A employing glucose substrate in extended culture operation. Substrate concentration, $S = 9.2 \text{ g l}^{-1}$, kept constant by substrate feed. Aeration rate: 0.55 vvm [16, 154].

upper part: dissolved oxygen concentration as a function of the cultivation time. Longitudinal position of the O_2-probes:

■ $z = -0$ (just below the aerator)
● $z = 0.09$ (at the aerator)
△ $z = 0.90$ (at the column head)

lower part: variation of the cultivation with time

■ glucose uptake rate (g l^{-1} h^{-1})
○ (dry) cell mass concentration, X, (g l^{-1})
△ oxygen uptake rate (g l^{-1} h^{-1})
▼ CO_2 production rate (g l^{-1} h^{-1})

column [218]. These measurements yielded k_L values for *H. polymorpha* which were by a factor of two higher than those for *E. coli* in the presence of Desmophen. In model media this antifoam effect is much larger (by a factor 5 to 6) [218].

Unexpected were the small d_s values in the presence of Desmophen. The large amount of protein (meat extract) in the *E. coli* medium seems to partly compensate the antifoam effect. This can also be recognized by means of a definite foam formation tendency in spite of the presence of Desmophen.

The courses of $k_L a^\alpha$ and a as a function of the cultivation time are rather different during *H. polymorpha* cultivations on ethanol and glucose substrate [16, 153, 154, 181]. When employing glucose, the dissolved oxygen saturation diminishes with increasing t as usual, but at 3.5 h passes a maximum and gradually increases (Fig. 58, upper part). This happens at the end of the exponential growth (Fig. 58 lower part). However the deviation from exponential growth was not caused by growth limitation, because sufficient substrate and dissolved oxygen were present (Figs. 58 and 59 upper part). The reduction of the growth rate is accompanied by diminutions of the yield coefficients $Y_{X/O}$ and $Y_{X/S}$ and the cell mass productions (Fig. 59 middle and lower parts). Also ethanol is produced (Fig. 59 upper part). Obviously

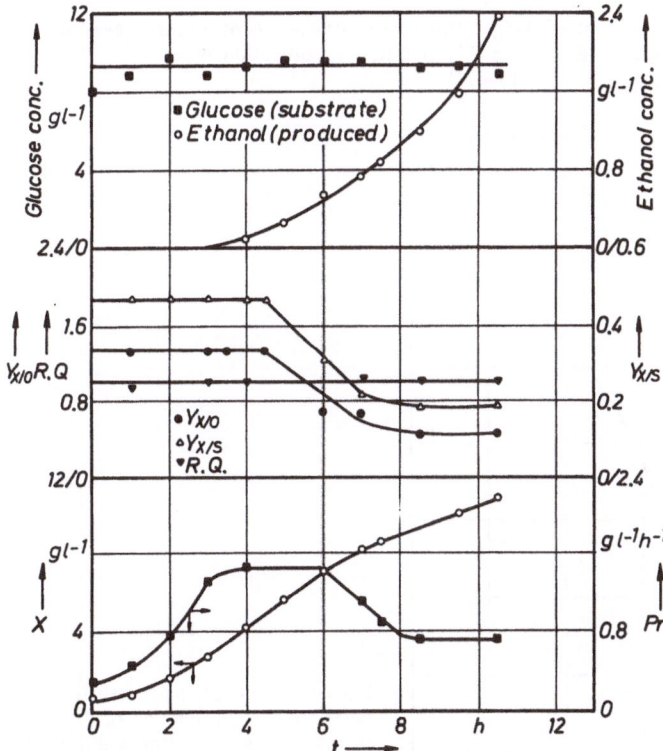

Fig. 59. Variation of yield coefficients, $Y_{X/O}$ and $Y_{X/S}$, respiratory quotient, R.Q., cell mass concentration, X, cell productivity, Pr, and produced ethanol concentration as a function of cultivation time in Reactor A during the cultivation of *H. polymorpha* at high glucose concentrations $(S = 9.2 \, g \, l^{-1})$ [16, 154]

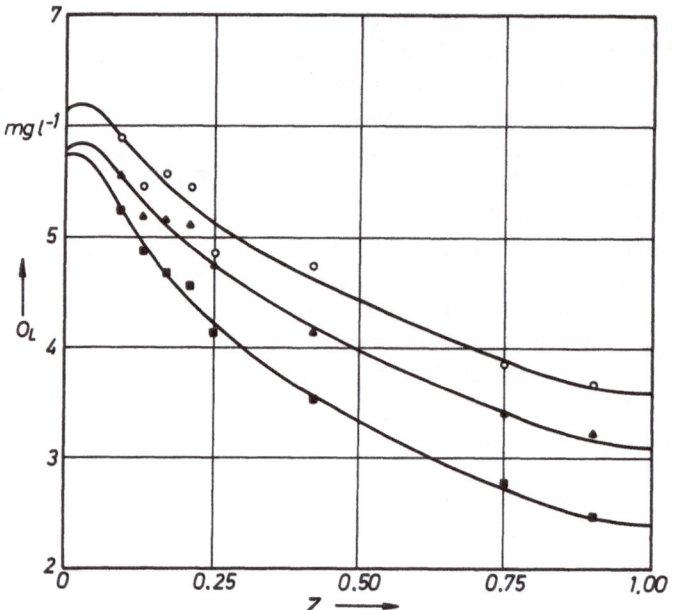

Fig. 60. Longitudinal profiles of dissolved oxygen (mg l^{-1}) in Reactor A during the cultivation of *H. polymorpha* on glucose substrate (S = 9.2 g l^{-1}) at different cultivation times, t [16, 154];

1 ■ t = 6.0 h	X = 7.10 g l^{-1}
2 ▲ t = 7.0 h	X = 9.65 g l^{-1}
3 ○ t = 9.5 h	X = 10.05 g l^{-1}

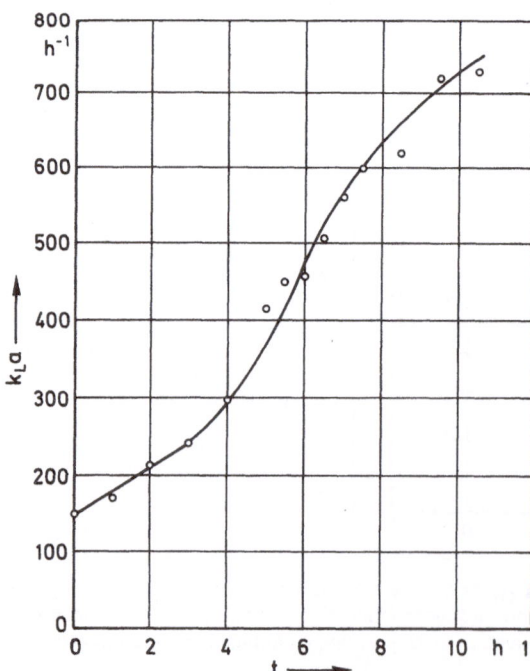

Fig. 61. Volumetric mass transfer coefficient, $k_L a$, as a function of the cultivation time, t, in reactor A during the cultivation of *H. polymorpha* employing glucose substrate, S = 9.2 g l^{-1}. Aeration rate: 0.5 vvm [16, 154]

some kind of repression has occurred due to glucose. (With the diminution of glucose concentration this effect can be reduced.) The longitudinal dissolved oxygen profiles are similar in this range to those measured in systems with ethanol substrate at the end of the exponential growth phase (Figs. 50 and 60). By fitting the calculated profiles to the measured ones the corresponding $k_L a^\alpha$ values were determined. Fig. 61 shows that $k_L a^\alpha$ increases with t. The same holds true for a (Fig. 62). This is caused by the increasing ethanol concentration. The k_L value is nearly constant (Fig. 63).

A comparison of $k_L a^\alpha$ values with ethanol and/or glucose substrate shows that the former $k_L a^\alpha$ is much higher ($1400 \rightarrow 800 \text{ h}^{-1}$) than the latter ($150 \rightarrow 700 \text{ h}^{-1}$) (Figs. 53 and 61). With increasing alcohol concentration in the glucose system the $k_L a^\alpha$ values also approach 800 h^{-1}. The same is true for the specific interfacial area, a, with ethanol: $a = 600 \rightarrow 1400 \text{ m}^{-1}$ (Fig. 54) and with glucose: $a = 200 \rightarrow 1000 \text{ m}^{-1}$ (Fig. 62). k_L values are also higher in the ethanol (Fig. 55) than in the glucose (Fig. 63) system but with increasing time and ethanol concentration they approach the same value ($20 \times 10^{-3} \text{ h}^{-1}$).

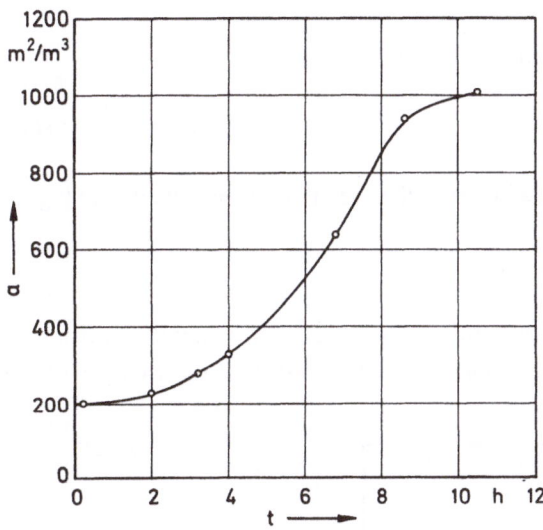

Fig. 62. Specific interfacial area, a, as a function of the cultivation time, t, in Reactor A during the cultivation of *H. polymorpha* employing glucose substrate, S = 9.2 g l^{-1}. Aeration rate: 0.5 vvm [16, 154]

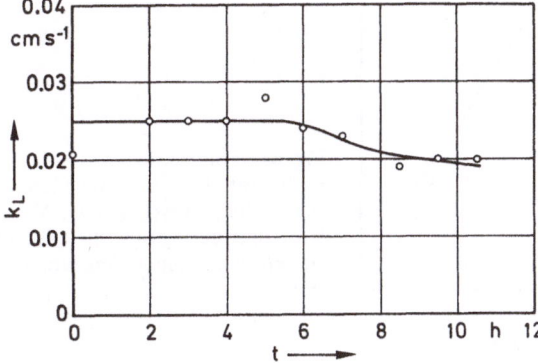

Fig. 63. Mass transfer coefficient, k$_L$, as a function of the cultivation time, t, in Reactor A during the cultivation of *H. polymorpha* on glucose substrate, S = 9.2 g l^{-1}. Aeration rate: 0.5 vvm [16, 154]

The influence of ethanol concentration on k_La^α is shown in Figs. 13 and 15. k_La^α increases with increasing ethanol concentration. The influence of the superficial gas velocity on OTR is shown in Fig. 14 and on k_La^α in Fig. 16. Both of them indicate that with increasing superficial gas velocity k_La^α increases, passes a maximum, then diminishes. This is due to the increasing medium recycling rate, $\dot{V}_R$. In Fig. 64 k_La is shown as a function of $\dot{V}_R$ during *E. coli* cultivation [17, 305]. The reduction of k_La with increasing $\dot{V}_R$ is significant. This is due to the reduction of E_G. According to Reith [304] the mean relative gas holdup, $(E_G)_{AL}$, can be calculated for liquid recycling by:

$$(E_G)_{AL} = \frac{w_{SG}}{2w_{SG} + w_S} \tag{122}$$

where w_S is the relative bubble swarm velocity with regard to the liquid velocity.
Since in bubble columns $(E_G)_{B.C.}$ is given by Eq. (123),

$$(E_G)_{B.C.} = \frac{w_{SG}}{w_S}, \tag{123}$$

the reduction of E_G due to air lift liquid recirculation is given by:

$$\frac{(E_G)_{AL}}{(E_G)_{B.C.}} = \frac{w_S}{2w_{SG} + w_S}. \tag{124}$$

This reduction of E_G is mainly caused by the increase of liquid velocity due to higher $\dot{V}_R$ (see Eq. (127)).

Substrate limited growth

In Fig. 65, the course of *H. polymorpha* cultivation on ethanol substrate is shown [16, 153, 282]: ethanol feed rate, $\dot{m}_{EtOH}$, ethanol concentration, S_{EtOH}, dissolved

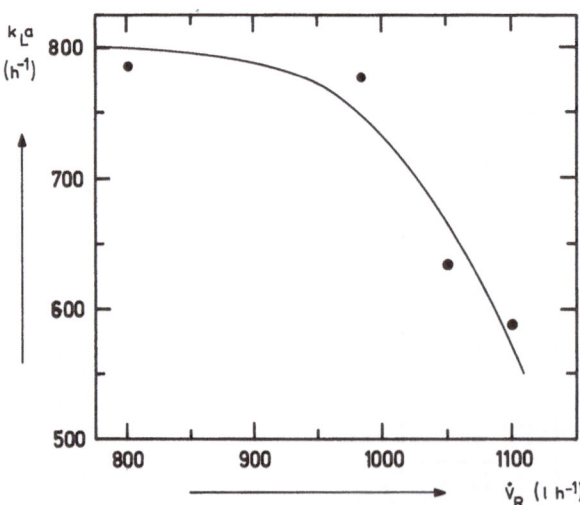

Fig. 64. Volumetric mass transfer coefficient, k_La, as a function of the medium recycling rate, $\dot{V}_R$ in Reactor A during the cultivation of *E. coli* in continuous operation [17, 305)]

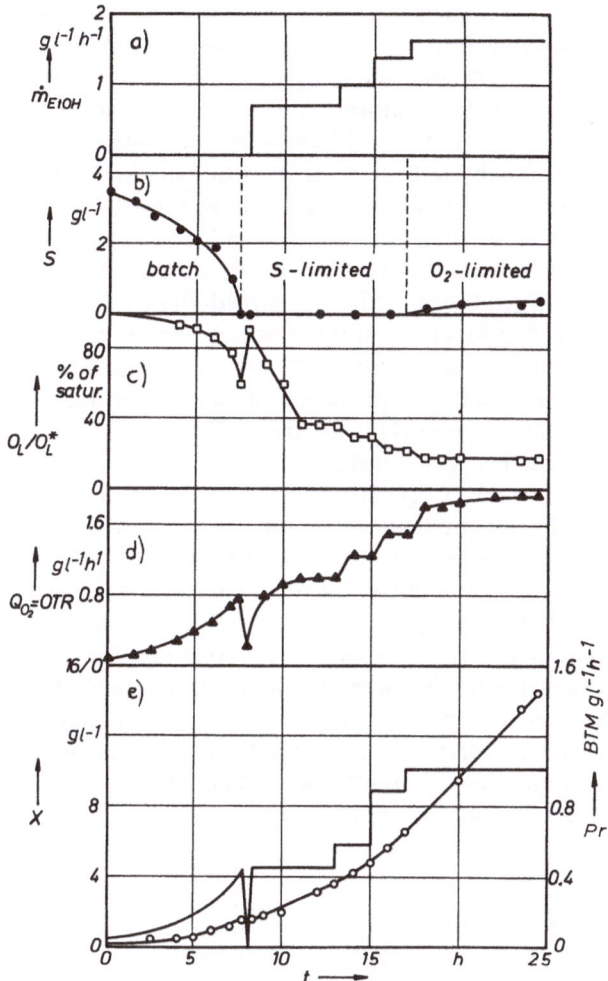

Fig. 65. Growth transitions during the cultivation of *H. polymorpha* in Reactor A employing ethanol substrate [16, 282].
a ethanol feed rate, $\dot{m}_{EtOH}$; b ethanol concentration, S; c relative dissolved oxygen saturation, DOS; OTR; e (dry) cell mass concentration, X, and cell productivity, Pr, as a function of the cultivation time

oxygen saturation, O_2-sat., OTR, Q_{O_2}, cell mass concentration, X, and cell mass productivity, Pr, are plotted as functions of the cultivation time, t. The cultivation started with batch mode until the substrate was consumed. Then the cells were fed batch cultivated at first at a low feed rate. The feed rate then was increased step by step. Substrate limited growth prevailed. By increasing the ethanol feed rate the substrate limited growth turned into oxygen transfer limited growth at a given feed rate and the ethanol concentration began to increase as can be seen from Fig. 65. At t = 7.5 h the substrate was consumed. Accordingly the cell growth stopped, the productivity diminished to zero, the OTR also reduced and the dissolved oxygen saturation increased. Shortly yfter t = 7.5 h the substrate feed was started at 0.71 g EtOH l^{-1} h^{-1}. The cells began to grow again, Pr and OTR increased dissolved oxygen saturation diminished, but the substrate concentration remained at a very low level. No ethanol could be detected. By increasing the substrate feed rate, the growth rate, Pr and OTR increased and dissolved oxygen saturation decreased. At every substrate feed rate a quasi-steady state value of these variables was attained. At t = 17 h the substrate feed rate was increased again, now from 1.37 to 1.63 g l^{-1} h^{-1}. As a result the substrate limited growth turned into oxygen transfer limited growth. The dissolved oxygen saturation was reduced to its critical value, O_L^c/O_L^*, below which oxygen could not be utilized by the organisms. Hence the maximum OTR was attained, which controlled the growth rate. Since under these conditions more substrate was fed than could be consumed, the ethanol concentration increased. The maximum productivity was attained and remained constant (1.05 g l^{-1} h^{-1}) even with increasing substrate concentration. To make the transition from substrate limited growth to oxygen transport limited growth more evident, the productivity Pr, and OTR are plotted as functions of the dissolved oxygen saturation in Fig. 66. For dissolved oxygen saturation larger than 20% substrate limited growth exists. Below this value oxygen transfer limitation prevails. One can also see how with increasing dissolved O_2 driving force (1 − O_L/O_L^*) Pr and OTR increase. At the boundary between substrate limited and oxygen transfer limited growth a productivity of 0.95 g l^{-1} h^{-1} can be attained. In the oxygen transfer limited range Pr = Pr$_m$ = 1.05 g l^{-1} h^{-1}. In Fig. 67a, the longitudinal dissolved oxygen concentration profiles, which correspond to the run represented by Fig. 65, are shown for different cultivation times, t. Curves 1 to 4 are characteristic for nonlimited growth, 5 and 6 for substrate limited growth and 7 for oxygen transfer limited growth. The curves in substrate limited growth show that the dissolved oxygen concentration in the upper half of the column increased after its decrease in the lower half of the column. This peculiar behaviour can be explained by the change of the nonlimited growth in the lower column half to a substrate limited one in the upper column half. Hence the oxygen uptake rate diminishes in the upper column half. Since the dissolved oxygen concentration at the interface remained the same, the necessary

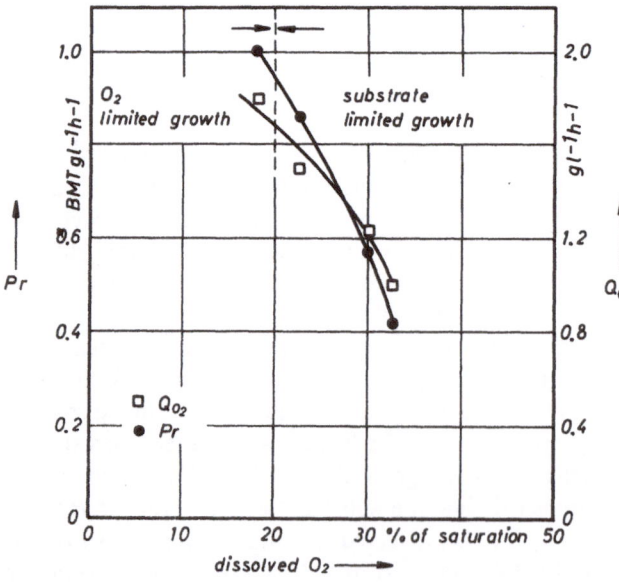

Fig. 66. *H. polymorpha* cell productivity and OTR in Reactor A as a function of the relative dissolved oxygen saturation, DOS, employing substrate ethanol [16, 282)]

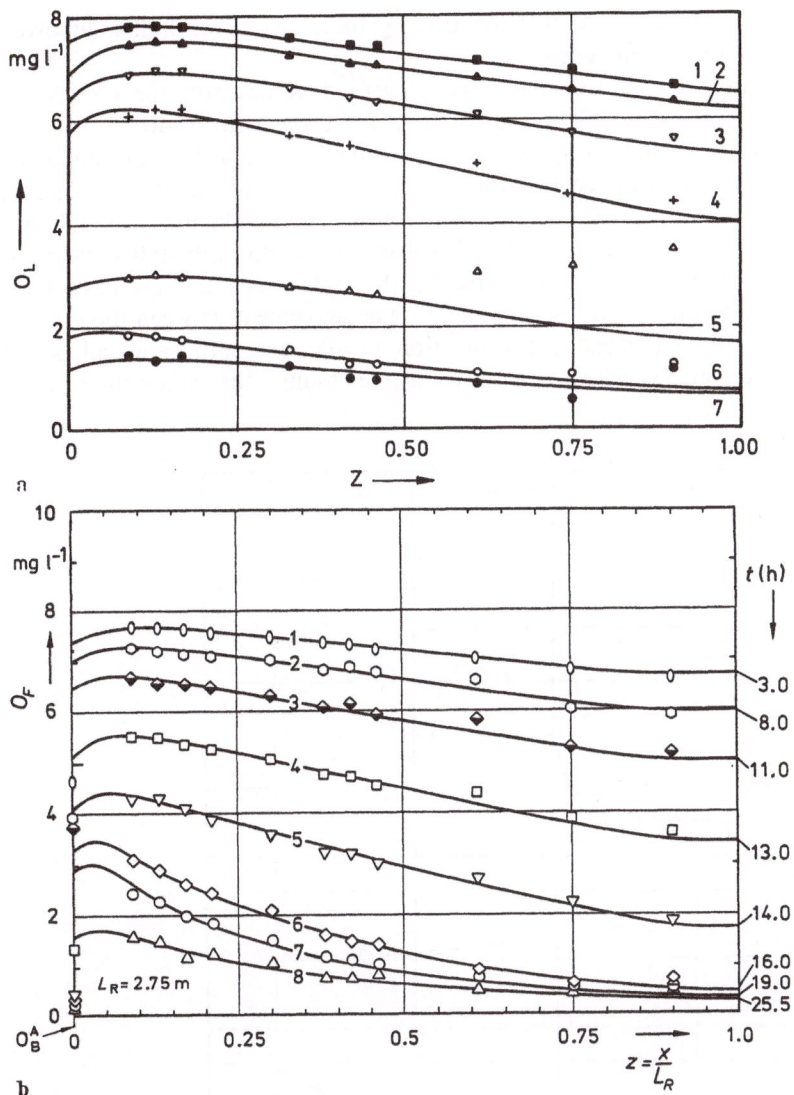

Fig. 67a. Longitudinal dissolved oxygen concentration profiles at different cultivation times in Reactor A during the cultivation of *H. polymorpha* on ethanol substrate. Profiles were calculated by assuming perfect mixing with regard to the substrate [16, 282].

■ *1* $t = 1.5$ h   △ *5* $t = 8.0$ h
▲ *2* $t = 4.0$ h   ○ *6* $t = 9.0$ h
▽ *3* $t = 6.0$ h   ● *7* $t = 12.0$ h
+ *4* $t = 7.0$ h

b. Longitudinal dissolved oxygen concentration profiles at different cultivation times in Reactor A during the cultivation of *H. polymorpha* on ethanol substrate. Profiles were calculated by taking the longitudinal concentration profile of substrate into account [181, 272].

○ *1* $t = 1.5$ h   ┐
○ *2* $t = 6.0$ h   │
□ *3* $t = 7.0$ h   ⎬ batch
○ *4* $t = 7.7$ h   │
◆ *5* $t = 8.0$ h   ┘

▽ *6* $t = 12.0$ h ⎫ substrate limited growth
◇ *7* $t = 14.0$ h ⎬
△ *8* $t = 17.0$ h ⎫ oxygen transfer limited growth
● *9* $t = 20.0$ h ⎬

oxygen could be supplied by a reduced driving force. As a result the dissolved oxygen saturation, O_L/O_L^*, increased.

The symbols in Fig. 67a represent the measured values and the curves the calculated profiles assuming CSTR behaviour in the tower with regard to the cells and substrate and a definite longitudinal dispersion with regard to the dissolved oxygen [181, 269–271]. One can see that these assumptions do not hold for substrate limited systems. These profiles can only be described by employing a model with distributed parameters with regard to the dissolved oxygen and substrate concentration and lumped parameters with regard to the cells. Such models have been developed by Luttmann [181] and Scheiding [272]. The agreement between the courses of longitudinal dissolved oxygen concentration profiles measured and calculated by these models are also excellent (Fig. 67b), if one assumes that along the tower a

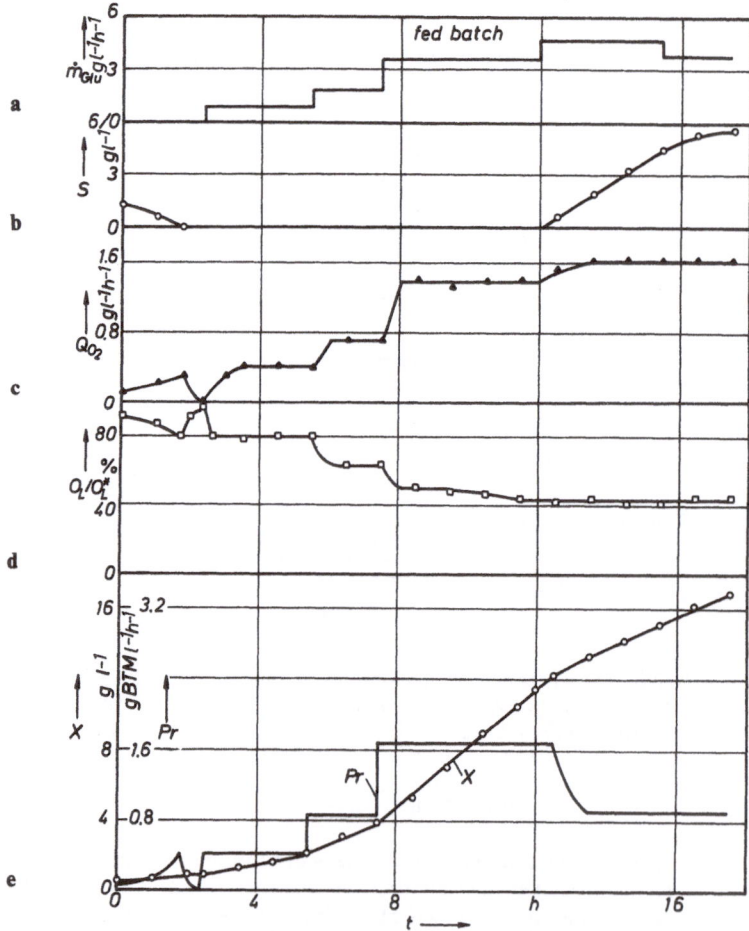

Fig. 68. Growth transitions during the cultivation of *H. polymorpha* in Reactor A on substrate glucose [16, 282]; **a** glucose feed rate, $\dot{m}_{Glu}$; **b** glucose concentration, S; **c** oxygen transfer rate, Q_{O_2}; **d** relative dissolved oxygen saturation, DOS; **e** (dry) cell mass concentration, X, and cell productivity, Pr; as functions of the cultivation time, t [16, 282].

change of nonlimited growth to substrate limited growth occurs. Analogous to the run plotted in Fig. 65, the glucose concentration was also varied in other runs [16, 153, 282] (Fig. 68). After batch cultivation and substrate consumption the substrate feed was started at a low level and then increased step by step. Again by consumption of the substrate the growth rate, the productivity and OTR were reduced to zero and dissolved oxygen saturation increased. By increasing the substrate feed rate step by step quasi-stationary states were maintained. Between $t = 2.5$ and 12 h no glucose could be detected in the medium. Thus, substrate limited growth prevailed. Up to this point no difference between the behaviour of the ethanol and glucose systems could be observed. By increasing the glucose feed rate from 3.6 to 4.7 $g\,l^{-1}\,h^{-1}$ the substrate concentration began to increase. Analogous to the ethanol system one would expect, from this point on, oxygen transfer limited growth. However, dissolved oxygen saturation did not change, the OTR increased only slightly, but its maximum had not been attained. All these data indicate that the cell had enough oxygen ($O_L/O_L^* \simeq 0.4$) and enough substrate, i.e. non limited growth existed. However, the productivity did not increase, but even decreased probably due to glucose repression, which is accompanied by ethanol production. The maximum cell productivity (1.7 $g\,l^{-1}\,h^{-1}$) was attained at a substrate feed rate of 3.6 $g\,l^{-1}\,h^{-1}$ under substrate limited conditions. The transition from substrate limited growth to oxygen transfer limited growth could not be achieved under these conditions.

In Fig. 69, the longitudinal profiles of the dissolved oxygen are again plotted for different cultivation time, t, by employing glucose as substrate. Curves 1 to 4

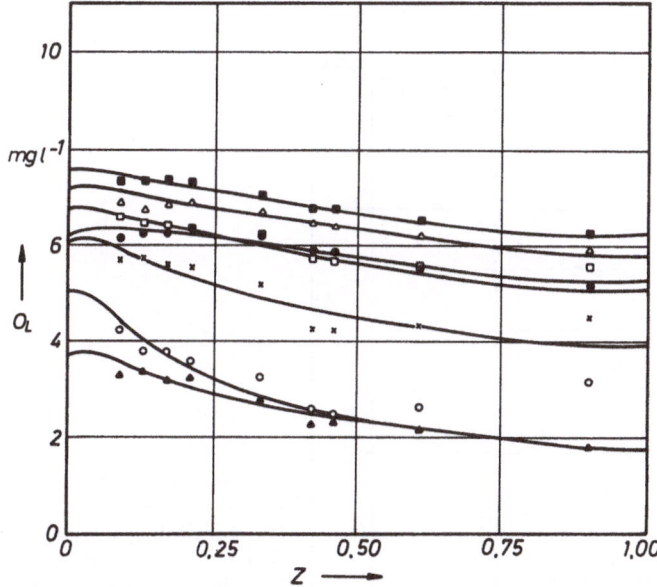

Fig. 69. Longitudinal dissolved oxygen concentration during the cultivation of *H. polymorpha* in Reactor A on substrate glucose at different cultivation times, t [16, 282].

1 t = 0.01 h	*5* t = 7.5 h
2 t = 1.0 h	*6* t = 8.5 h
3 t = 1.75 h	*7* t = 14.5 h
4 t = 5.5 h	

are characteristic for nonlimited growth, 5 and 6 for substrate limited growth and curve 7 for glucose repression. Analogous to the ethanol system in the nonlimited growth region, the dissolved oxygen concentration diminishes in the lower half of the tower and increases in the substrate limited growth region in the upper half of the tower. These longitudinal dissolved oxygen concentration profiles could also be described by a distributed parameter model with regard to the substrate and oxygen concentration.

Figure 19 shows how $k_L a^\alpha$ varies during the run represented by Fig. 65. After its initial increase $k_L a^\alpha$ is reduced with diminishing ethanol concentration from 1600 to 220 h^{-1}. The latter prevailed in the substrate limited range. In the oxygen transfer limited range, and at detectable ethanol concentrations, $k_L a^\alpha$ increases again and approaches a constant value (350 h^{-1}). The course of a is similar to that of $k_L a^\alpha$.

In Fig. 70, $k_L a^\alpha$ and a are shown as a function of the cultivation time for the run represented by Fig. 68. As one can see, no dramatic variations of $k_L a^\alpha$ and a can be observed.

The gradual increase in $k_L a^\alpha$ and a for $t > 4$ h is due to increasing alcohol concentration.

It is interesting to consider how the variation of the medium recirculation rate, $\dot{V}_R$, influences the longitudinal concentration profiles of the medium components in the tower. Typical cultivation conditions by employing *E. coli* in a continuously operated tower loop reactor at high medium recirculation rates, $\dot{V}_R$, are shown in Figs. 71 and 72 [17, 305]. The steady state is quickly attained at high $\dot{V}_R$. A substrate

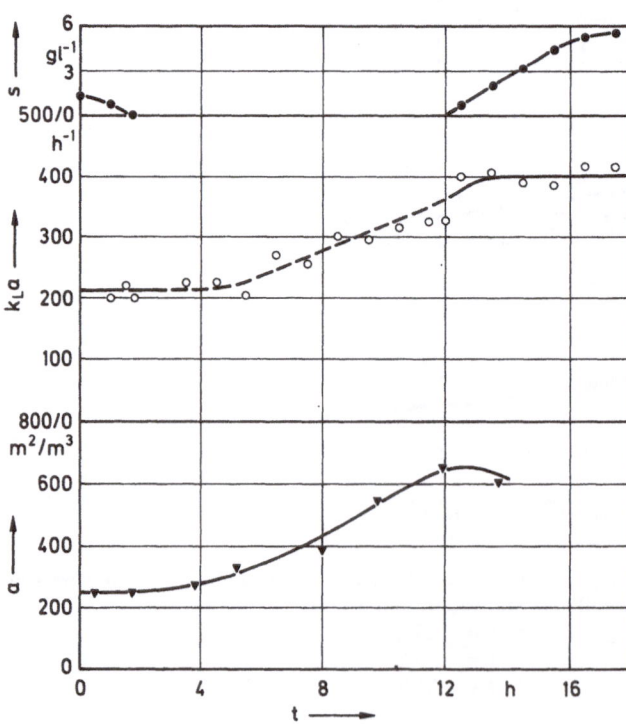

Fig. 70. Substrate concentration, S, volumetric mass transfer coefficient, $k_L a$, and specific interfacial area, a, during the cultivation of *H. polymorpha* in Reactor A on substrate glucose as function of the cultivation time, t [16, 282]; ● C_{Glu}; ○ $k_L a$; ▲ a

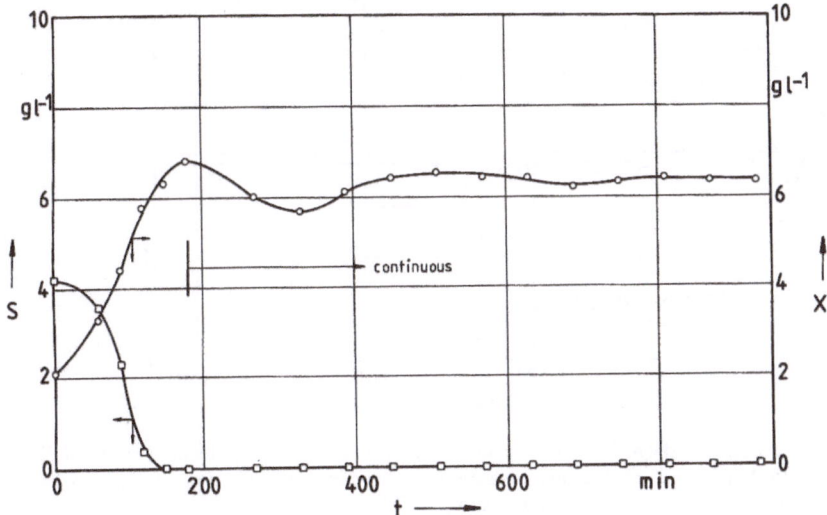

Fig. 71. Cell mass concentration, X, and substrate concentration, S, during the cultivation of *E. coli* in Reactor A in batch and continuous operation as functions of the time, t. $S_0 = 5.5\,g\,l^{-1}$, $D = 0.44\,h^{-1}$, $\dot{V}_R = 1400\,l\,h^{-1}$, $\gamma = 70$ [17, 305]

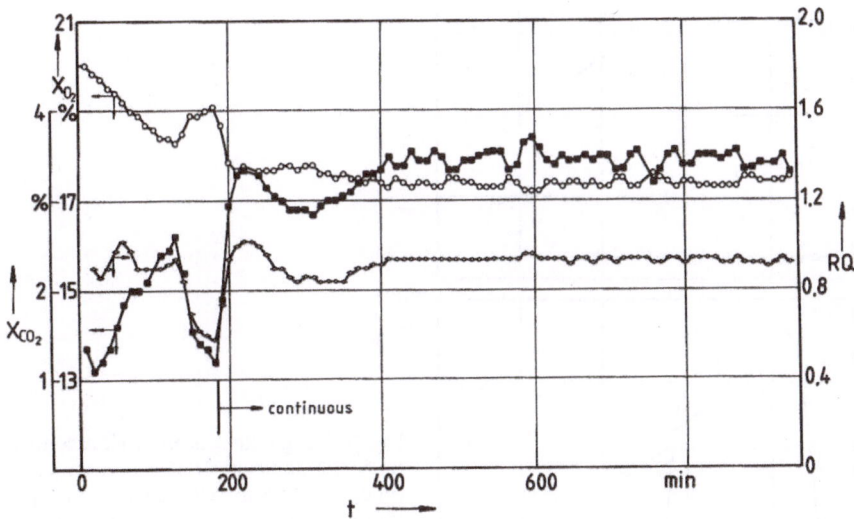

Fig. 72. Oxygen and CO_2 concentrations, x_{O_2} and x_{CO_2}, in gas phase and respiratory quotient, R.Q., during the cultivation of *E. coli* in Reactor A in batch and continuous operation as functions of the time, t. (For operation conditions see Fig. 71) [17, 305], $\circ$ x_{O_2}, $\blacksquare$ x_{CO_2}, $\diamond$ R.Q.

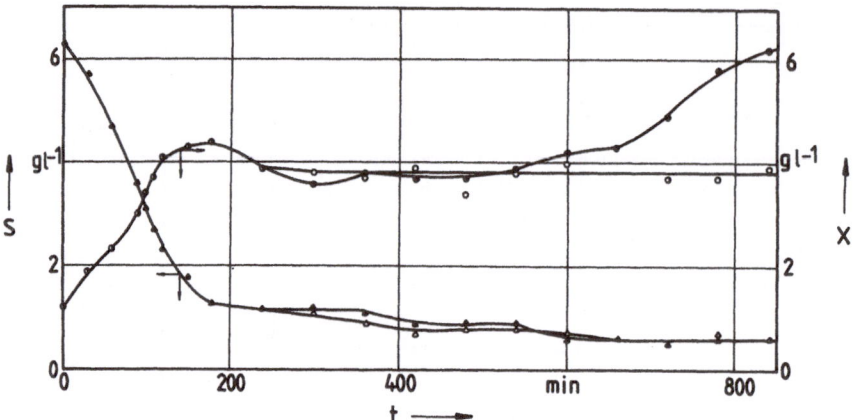

Fig. 73. Cell mass concentration, X, and substrate concentration, S, during the cultivation of *E. coli* in Reactor A in batch and continuous operation as functions of the time, t. $S_0 = 5.4 \, g \, l^{-1}$, $D = 0.37 \, h^{-1}$, $\dot{V}_R = 0$, $\gamma = 0$ (no medium recycling) [17, 305]:

● X at x = 25 cm   ▲ S at x = 25 cm
○ X in loop   △ S in loop

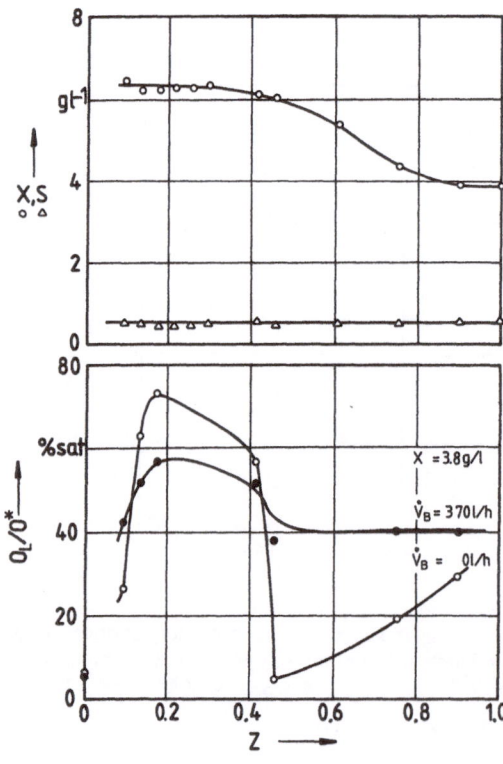

Fig. 74. Longitudinal profiles of cell concentration, X, and substrate concentration, S (in upper part) as well as dissolved oxygen saturation, DOS, during the cultivation of *E. coli* in Reactor A in steady state operation [17, 305]. ○ X, △ S, ○ DOS. (For operation conditions see Fig. 73); ● DOS at $\dot{V}_R = 370 \, l \, h^{-1}$

conversion, U_s, of ca. 100% is achieved at $D = 0.44\,h^{-1}$. On the other hand, in Figs. 73 and 74 cultivation conditions are shown in the absence of medium recirculation. During the batch phase the medium was recirculated: $\dot{V}_R = 375\,l\,h^{-1}$ ($\gamma = 22.2$). After starting the medium feed: $D = 0.37\,h^{-1}$, in continuous operation, $\dot{V}_R$ was reduced to zero. During the first 300 min of continuous operation the process variables are similar to those measured at high $\dot{V}_R$ values. However, the cell mass concentration, X, which was measured at $x = 25$ cm (above the aerator), gradually increased (Fig. 73) due to the nonuniform longitudinal profile of cell mass concentration (Fig. 73). As one can see from this figure, which is characteristic for the end of this run, the cell mass concentration is 50% higher in the lower half of the tower than in the upper half. In spite of this the substrate concentration remained uniform in the tower. The longitudinal concentration profiles of dissolved oxygen are considerably nonuniform. In the lower half of the tower high dissolved oxygen concentrations prevail (Fig. 74). They drop to very low values at about the half height of the tower and then gradually increase. When increasing the recirculation rate, $\dot{V}_R$, this profile becomes more uniform (Fig. 74). In the absence of medium recirculation, cell sedimentation occurs, which leads to oxygen transfer limitation in the upper half of the tower. This causes cell flocculation, which increases cell sedimentation. Thus sedimentation and flocculation are mutually amplified. Under these conditions the substrate conversion reduces to 88.5% and the yield coefficient, $Y_{X/S}$, to 0.8.

7.2.2 Ten Stage Concurrent Tower Loop Reactor (Reactor B) [17, 315]

Figures 75 and 76 show a typical run in the ten stage tower loop reactor by cultivating *E. coli* with $D = 0.33\,h^{-1}$ and $\dot{V}_R = 233\,l\,h^{-1}$. The courses of X and S are similar to those in the single stage tower by employing the same bacterium (Fig. 71), but in the ten-stage tower the steady state has not yet been achieved up to $t = 800$ min. The variation of X with t is not caused by the variation of the local cell mass concentrations (X's in the tower and in the loop are constant and equal), but

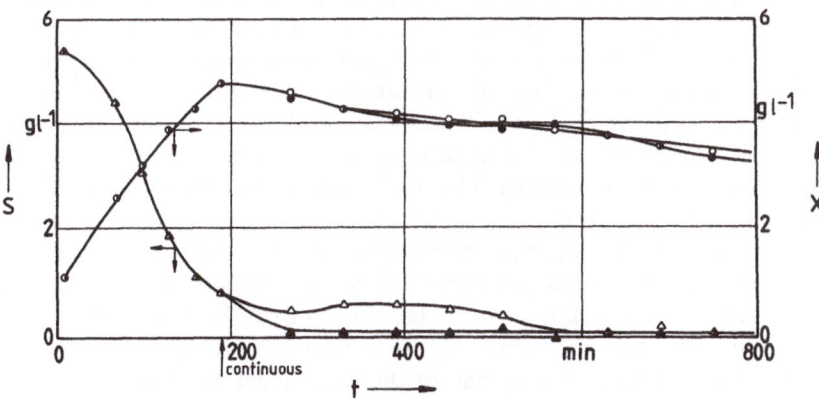

Fig. 75. Cell mass concentration, X, and substrate concentration, S, during the cultivation of *E. coli* in Reactor B (ten-stage tower loop) in batch and continuous operations as functions of the time, t. $S_0 = 5.6\,g\,l^{-1}, D = 0.33\,h^{-1}, \dot{V}_R = 233\,l\,h^{-1}$ [17, 305];
- ● X at x = 25 cm
- ○ X in loop
- ▲ S at x = 25 cm
- △ S in loop

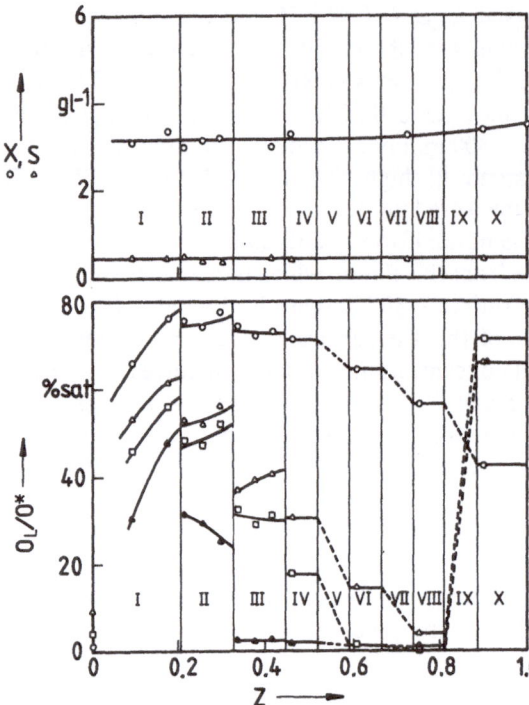

Fig. 76. Longitudinal profiles of cell mass concentration, X, substrate concentration, S, (upper part) during the cultivation of *E. coli* in Reactor B in batch and continuous operation [17, 305];

o X,   S (continuous) upper part
o DOS at t =   40 min (batch)
▲ DOS at t =   220 min (conti.)
                                     lower part
□ DOS at t =   270 min (conti.)
▲ DOS at t = 400−820 min (conti.).
I—X are the stages counted from the bottom

by the variation of the overall cell mass concentration in the tower. The differences in S in tower and loop disappear after 600 min (Fig. 75). The uniform cell mass concentrations, X, and substrate concentrations, S, in the ten-stage tower are shown in Fig. 76. On the other hand, the longitudinal dissolved oxygen saturation profiles are considerably nonuniform. At high medium recirculation rate and batch operation the dissolved oxygen saturation diminishes with increasing stage number (counted from the bottom), but it remains higher than 0.5. Thus no oxygen transfer limitation occurs. In continuous cultivation the dissolved oxygen saturation drops to zero. With increasing cultivation time the numbers of stages increase in which the oxygen supply to the bacteria is not sufficient (Fig. 76). (The increase of the oxygen concentration in the 10th stage is caused by the mechanical foam destroyer at the top of the column, which also acts as an aerator). The inadequate oxygen transfer rate in the ten-stage tower is due to the formation of liquid free layers below the perforated trays, which acted as very ineffective gas distributors, because of their large (3 mm) hole diameters and free cross sectional area (6.52%). In addition, the efficiency of the stage separating trays probably deteriorated due to foam formation below the trays. The foam lamellae block the tray openings. Thus gas pressure increases. At a critical overpressure, the lamellae in one of the holes burst and the gas escapes through this hole of the tray to the next upper stage. Thus the tray actually acts as a single hole aerator plate which is especially ineffective.

When reducing the medium recirculation rate the cells are enriched in the 1st stage. However, X is constant in the rest of the stages. In the tower no cell sedimentation occurs. The high X-values in the 1st stage are caused by cell sedimentation in the loop. The substrate concentration is nearly constant in the loop system. The longitudinal profile of the dissolved oxygen saturation is similar to that found at higher medium recirculation rate [17, 305]. At a very low medium recircula-tion rate ($\dot{V}_R = 16\,l\,h^{-1}$, $\gamma = 1$) differences appear between X's and S's in tower and loop during batch cultivation (Fig. 77). At the beginning of the continuous culture, X is higher in the loop than in the 1st stage of the tower. With increasing cultivation time X in the loop gradually diminishes and in the 1st stage remains constant. After 550 min, X becomes higher in the 1st stage than in the loop (Fig. 77). During the crossing of these profiles and thereafter S increases considerably. However, S is always higher in the 1st stage than in the loop, because the feed is introduced into the 1st stage. It is not known why this crossing of X profiles occurs. No dramatic changes of x_{O_2}-, x_{CO_2}- and RQ-courses can be recognized during the X profile-crossing (Fig. 78). The longitudinal profiles of X, S and pH indicate (Fig. 79) that X has a slight minimum in the 2nd stage, and then slightly diminishes with increasing stage number. In the ten-stage tower loop the diminution of $\dot{V}_R$ also reduces the substrate conversion, U_S, and yield coefficient, $Y_{X/S}$.

7.2.3 Single-stage Countercurrent Tower Loop Reactor (Reactor C) [87, 88, 89]

H. polymorpha was cultivated in reactor C in extended culture under fixed operation conditions: the liquid volume of the cultivation medium was 50 l, the aerated height 160 cm, the aeration rate 1 vvm ($w_{SG} = 3.02$ cm s^{-1}), the liquid recirculation rate, $\dot{V}_R = 200\,l\,h^{-1}$ and ethanol concentration S = 1 % and/or 0.8 % per volume. At first antifoam free systems are considered. In Fig. 80, dissolved oxygen saturations in the middle of the tower and in the loop are shown as functions of cultivation time. The dissolved oxygen saturation in the loop is always much lower than in the tower. After 8 h no dissolved oxygen could be detected in the loop. Around this time the oxygen transfer limitation became effective and the specific growth rate

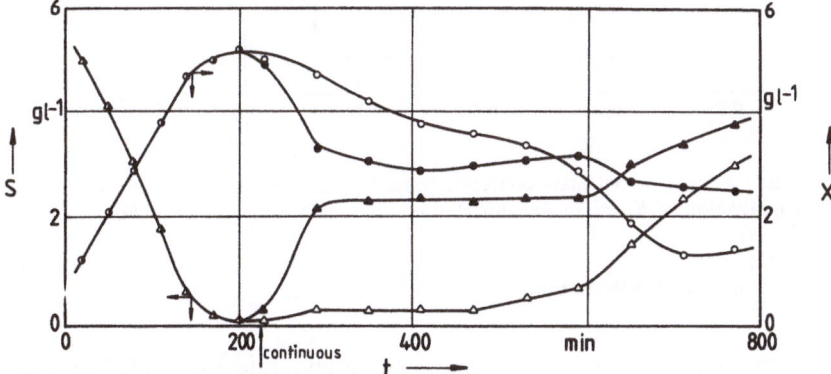

Fig. 77. Cell mass concentration, X, and substrate concentration, S, during the cultivation of *E. coli* in Reactor B in batch and continuous operations as function of the time, t. $S_0 = 5.8$ g l^{-1}, $D = 0.36$ h^{-1}, $\dot{V}_R = 16\,l\,h^{-1}$, $\gamma = 0.95$ [17, 305].
● X at x = 25 cm   ▲ S at x = 25 cm
○ X in loop   △ S in loop

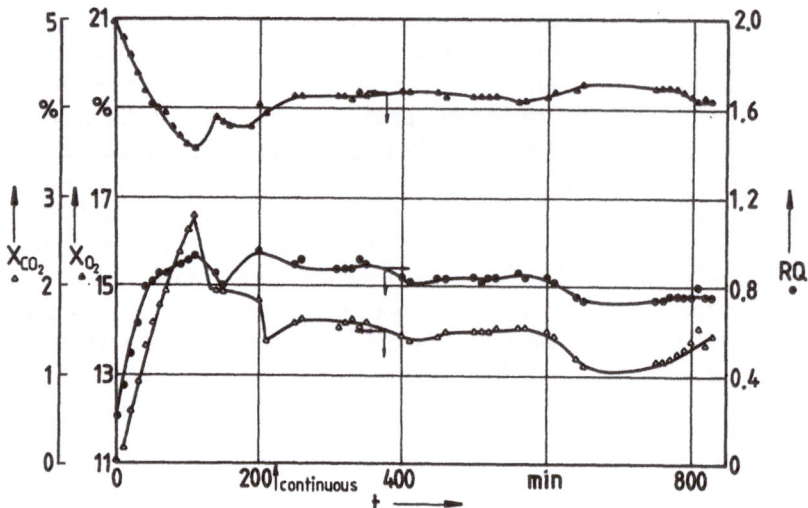

Fig. 78. Oxygen and CO_2 concentrations, x_{O_2} and x_{CO_2} in gas phase and respiratory quotient, R.Q., during the cultivation of *E. coli* in Reactor B in batch and continuous operation as function of the time, t [17, 305]. (For operation conditions see Fig. 77);
▲ O_2, △ CO_2, ● R.Q.

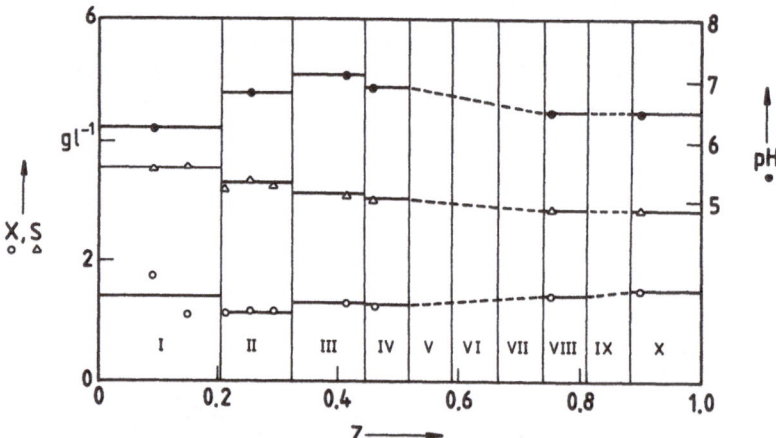

Fig. 79. Longitudinal profiles of cell mass concentrations, X, substrate concentration, S, and pH⸜ value during the cultivation of *E. coli* in Reactor B in continuous operation [17, 305]. (For operating conditions see Fig. 77);
○ X, △ S, ● pH;
I—X are the stages counted from the bottom

began to decrease. Also the curves of X, substrate uptake rate and OTR began to deviate from the exponential course and evened off (Fig. 81). This transition was completed after 12 h and the growth became linear in the entire reactor due to oxygen transfer limitation of growth.

A linear plot of X as a function of t indicates linear growth of the cells during this phase. At this time maximum cell productivity of 0.71 g l^{-1} h^{-1} and a maximum

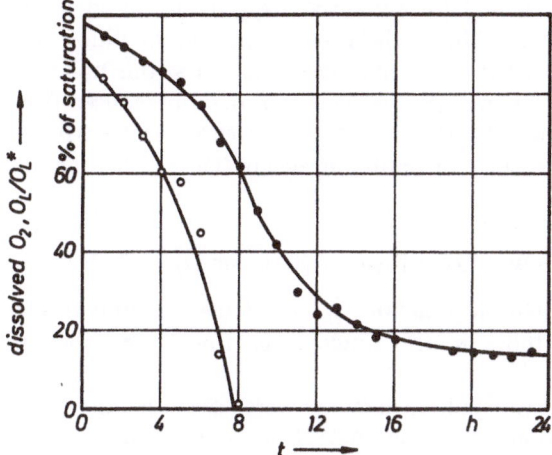

Fig. 80. Dissolved oxygen concentration as a function of the cultivation time, t, during the cultivation of *H. polymorpha* in Reactor C in absence of antifoam agent and employing substrate ethanol. (For operation conditions see Fig. 81) [87, 88];
o loop; • in the middle of the tower

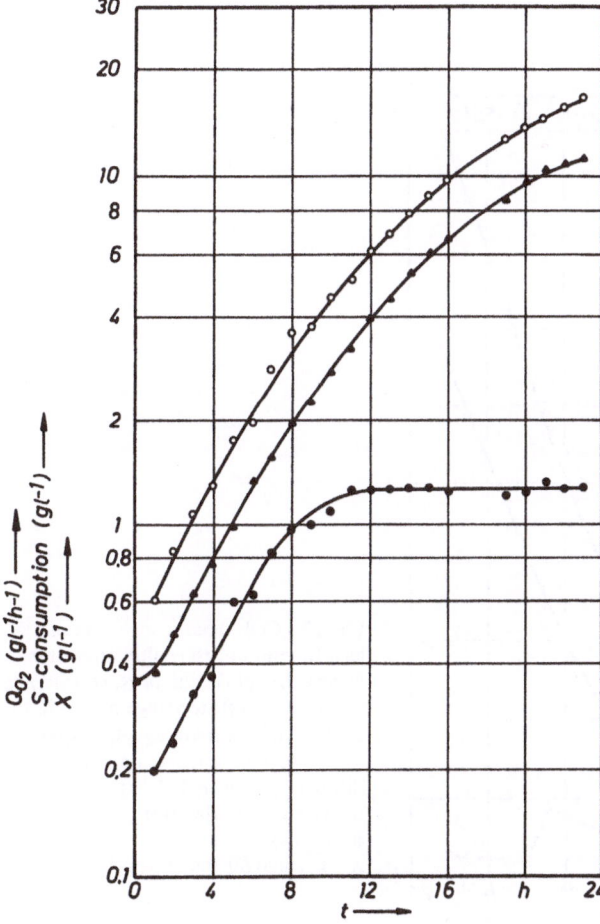

Fig. 81. Extended cultivation of *H. polymorpha* in Reactor C in absence of antifoam agent employing substrate ethanol $S = 1\%$, aeration rate: 1 vvm, medium recycling rate: $\dot{V}_R = 200 \, 1 \, h^{-1}$ [87, 88];
▲ cell mass concentration $(g \, l^{-1})$
● OTR $(g \, l^{-1} \, h^{-1})$
o consumed ethanol $(g \, l^{-1})$

OTR of $1.28 \text{ g l}^{-1} \text{ h}^{-1}$ were attained and the dissolved oxygen saturation in the middle of the tower amounted to 0.25. Under the given operation conditions the maximum productivity, controlled by the OTR, had been attained at this point.

In the presence of antifoam agent Desmophen, μ_m was considerably reduced from 0.25 to 0.21 h^{-1}, k_La from 250 to 160 h^{-1} and $Y_{X/O}$ increased from 0.55 to 0.69. The results obtained in the presence of antifoam agents (Desmophen and/or soy oil) are discussed in 7.2.4.

7.2.4 Three-stage Countercurrent Tower Loop Reactor (Reactor D) [87, 88, 89]

Again *H. polymorpha* was cultivated in Reactor D in extended culture under identical operation conditions to those in the single stage tower, described in 7.2.3.

In Reactor D perforated plates 0.5 mm in hole diameter were used as compartment separating trays. For the investigations in the absence of antifoam agents a 40 cm bubbling layer height was used which was controlled by the overflow. The basic problem in operating multistage bioreactors in the absence of antifoam agents had already appeared shortly after the inoculation. The free volume between the bubbling layer and the tray above the layer were filled with foam. By microflotation effect

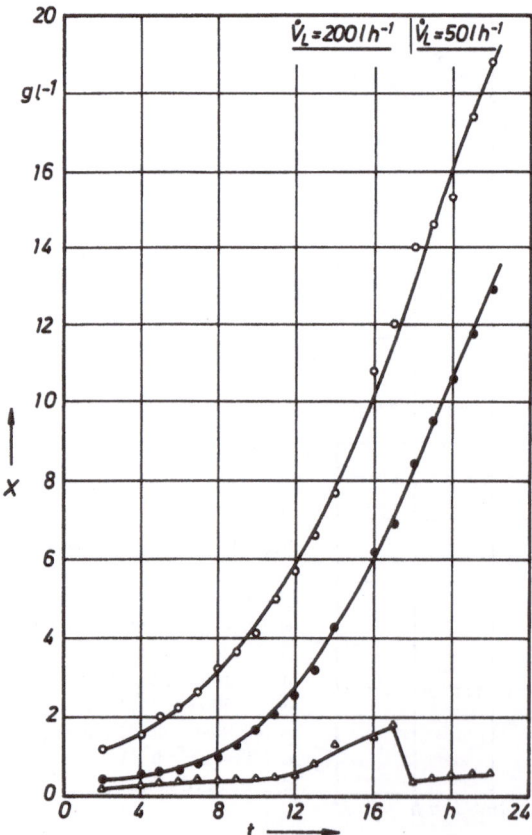

Fig. 82. Cell mass concentration, X, as a function of the cultivation time, t, during the cultivation of *H. polymorpha* in Reactor D (three-stage tower) in the absence of antifoam agents employing substrate ethanol (for operation conditions see Fig. 81) [87, 88];

o 1. stage (at the top)
● 2. stage
△ 3. stage (at the bottom)

(see Chapter 4.2.8, 6.4 as well as [247]), the cells, which were enriched in the foam, passed through the tray into the next upper stage. Here they again were enriched in the foam and so on. Through this microflotation effect the cells were enriched in the upper stage; in the lower stages the cell concentration was considerably diminished as can be seen from Fig. 82. It was not possible to reduce this cell segregation by increasing the liquid circulation rate since the liquid transport capacity of the overflow was considerably reduced when foaming medium was present. The same flotation effect was also found by Kitai et al. [47,48] who also employed a (concurrent) multistage tower. To avoid this cell segregation they employed antifoam agent. From Fig. 82 the influence of this flotation effect can clearly be recognized. In the highest stage the cell concentration was highest and in the lowest one lowest. At $t = 17.5$ h the liquid recirculation rate had to be reduced, since the transport capacity of the overflow tubes were no longer sufficient to transport the foaming medium from the upper to the lower stages. This reduction of liquid circulation rate to 50 l h^{-1} further reduced the cell concentration in the lower stage to very low values (0.5 g l^{-1}). Due to these different cell concentrations, the dissolved oxygen saturations were very different in the three stages: in the upper stage the lowest and in the lowest stage the highest dissolved oxygen saturations were attained. In the loop a high dissolved oxygen concentration could be preserved due to the high dissolved oxygen concentration and low cell mass concentration in the lowest stage, from which the medium was recycled. A comparison of X and O_L/O_L^* in the 1st (upper) and 2nd (middle) stages with those in the single stage tower (7.2.3) indicates that at the same X, the O_L/O_L^* in the multistage tower was higher than in the single stage one. Thus the OTR was higher in the multistage tower than in the single stage one. This is also shown by comparison of the volumetric transfer coefficients, $k_L a$'s, evaluated in single and multistage towers (Fig. 83). At long cultivation times the $k_L a$-value in the three stage tower was about four times as high as that in the single stage tower. This yielded a higher maximum OTR (2.20 g l^{-1} h^{-1}) in the three-stage tower than the one (1.28 g l^{-1} h^{-1}) in the single stage tower.

The considerable increase of $k_L a$ in the three-stage tower compared with the single-stage tower (Fig. 83) yields only a slight increase of the maximum cell productivity

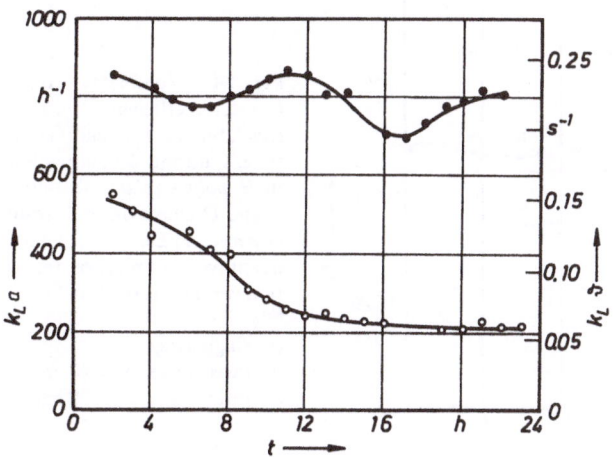

Fig. 83. Comparison of the volumetric mass transfer coefficients, $k_L a$'s in Reactor C (single stage tower loop) and Reactor C and D (single- and three-stage tower loops) [87,88];
o single-stage
● three-stage (h = 40 cm)

from 0.71 (in single-stage) to 0.80 g l^{-1} h^{-1} (in three-stage column). This is due to cell enrichment in the foam. Cell growth in the foam is possibly controlled by the substrate concentration and not by the dissolved oxygen concentration.

To eliminate the foam two antifoam agents were employed alternatively: Desmophen and soy oil [87, 88, 89].

The extended cultivation of *H. polymorpha* was carried out under the operation conditions specified in 7.2.3 and additionally employing an antifoam (*Desmophen 3600*) feed rate of 3 ml h^{-1}. To test the influence of the bubbling layer height, h, on the OTR and cell productivity, Pr, h = 40 and 20 cm were employed in the three-stage tower. The single-stage tower was operated with H = 160 cm. During the cultivations no foam formation was observed. Also no differences were found in the cell mass concentrations, X, of the three different stages.

In Fig. 84 the volumetric mass transfer coefficients, k$_L$a's, are plotted as functions of the cultivation time, t, for the single-stage as well as for the three-stage columns at h = 40 and 20 cm. One clearly recognizes that the lowest k$_L$a values were found in the single-stage tower and the highest ones in the three-stage tower with h = 20 cm. The installation of trays improves the k$_L$a values. This improvement increases with diminishing bubbling layer height, h. In all these systems, k$_L$a first increases and then passes a maximum at X $\simeq$ 3 g l^{-1} (Fig. 84), similar to the k$_L$a-behaviour found in concurrent tower loop reactors (7.2.1). This increase in k$_L$a after inoculation is probably caused by surface active compounds secreted by the cells. The use of antifoam agents reduces k$_L$a considerably as can be seen by comparing Fig. 83 with Fig. 84. This holds true for the single-stage as well as for the three-stage tower. With increasing cultivation time the oxygen uptake rate increases, the dissolved oxygen saturation diminishes and OTR increases. Again the highest OTR's were attained in the three-stage tower with h = 20 cm, and the lowest employing the single-stage tower, as expected. The same is true for the cell productivity: Pr = 0.69 g l^{-1} h^{-1} (single-stage), Pr = 0.75 g l^{-1} h^{-1} (three-stage, h = 40 cm), Pr = 1.09 g l^{-1} h^{-1} (three-stage, h = 20 cm).

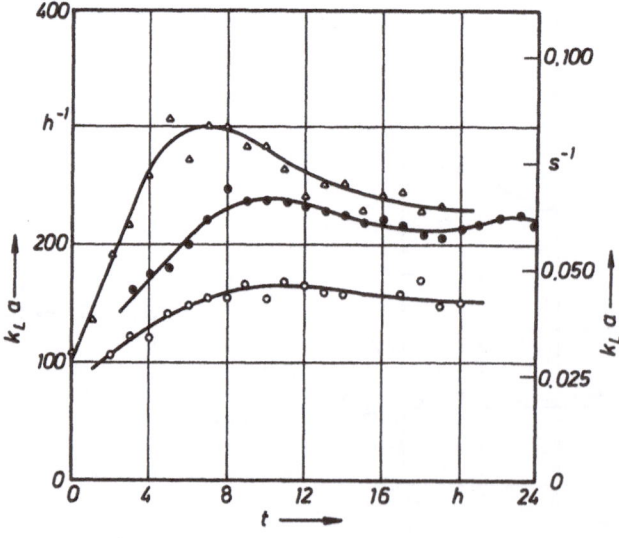

Fig. 84. Volumetric mass transfer coefficients, k$_L$a's, as functions of the cultivation time, t, during the cultivation of *H. polymorpha* in Reactors C and D employing substrate ethanol and Desmophen 3600 as antifoam agent. (For operation conditions see Fig. 81) [87, 88],

o   single-stage
●   three-stage, h = 40 cm
△   three-stage, h = 20 cm

The operation conditions were again identical with those given in 7.2.3, when *soy oil* with a feed rate of 0.5 ml h⁻¹ was employed as antifoam agent [87, 88, 89]. The soy oil was a less effective antifoam agent than Desmophen 3600. When employing soy oil the foam did not disappear, only its structure changed: the bubble size increased and the foam became less stable. After 8 h of cultivation, the cells were enriched again in the upper stage due to their microflotation (see Chapters 4.2.8 and 6.4 as well as [247]) However, the nonuniformity of the cell concentration was considerably less than for the antifoam-free system, the cell concentration in the upper stage was, at its maximum, by a factor of two higher than the one in the lowest stage. To reduce the foam amount at t = 8 h the antifoam feed rate was considerably increased for a short period of time and then decreased as can be seen from the upper part of Fig. 85. In the lower part of this figure the cell mass concentration, X, OTR, and the ethanol uptake rate are plotted as a function of the cultivation time, t. After inoculation the cells began to grow without a lag phase. In this exponential growth phase $\mu_m = 0.11$ h⁻¹ prevailed. During the time t = 6 to 9 h a reduction of μ_m occurred. After this period a new exponential growth phase began with $\mu_m = 0.26$ h⁻¹. Obviously the cells adapted themselves to the soy oil and used it as a substrate.

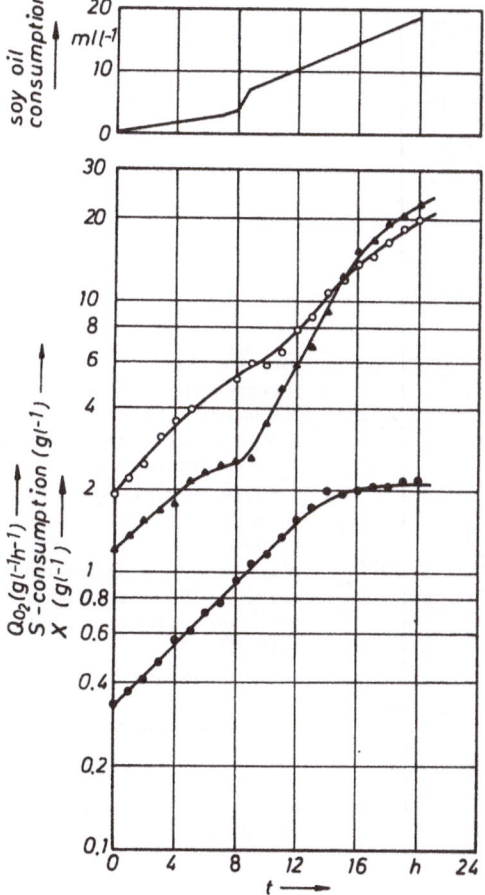

Fig. 85. Cultivation of *H. polymorpha* in Reactor D-40 (three-stage with h = 40 cm) employing substrate ethanol and soy oil as antifoam agent. (For operation conditions see Fig. 81) [87, 88],

• OTR, Q_{O_2} (g l⁻¹ h⁻¹)
o ethanol consumption (g l⁻¹)
▲ (dry) cell mass concentration (g l⁻¹)
top: soy oil added to the system (ml l⁻¹)

At t = 16 h the oxygen transport limited growth phase began (Fig. 85). The phenomenon of the two-stage growth with a lag phase (diauxie) is well known (e.g. [298, 306, 307]).

The ethanol uptake rate in Fig. 85 indicates that after t = 8 h its rate diminishes, although the specific growth rate increases. This clearly shows that the cells, after a short induction period, adapted themselves to the soy oil, after which they used two substrates. It should be mentioned that the OTR is not influenced by this change. The increase of the growth rate must be coupled with the increase in the oxygen yield coefficient, $Y_{X/O}$.

The volumetric mass transfer coefficients, $k_L a$'s, are plotted in Fig. 86 as functions of the cultivation time, t, for the single-stage tower loop (reactor C) and three-stage tower loop (reactor D) with h = 40 and 20 cm. In the upper part of this figure the soy oil feed rates are given as a function of t. One can see that up to t = 8 h nearly the same antifoam feed rate was employed for all runs. During this time, $k_L a$ in the three-stage tower with h = 20 cm was the largest and with h = 40 cm the smallest. The $k_L a$ in the single-stage was intermediate. For this unusual behaviour no satisfactory explanation can yet be given. After increasing the soy oil feed rates, the $k_L a$'s in the three-stage tower with h = 20 cm were highest and in the single-stage

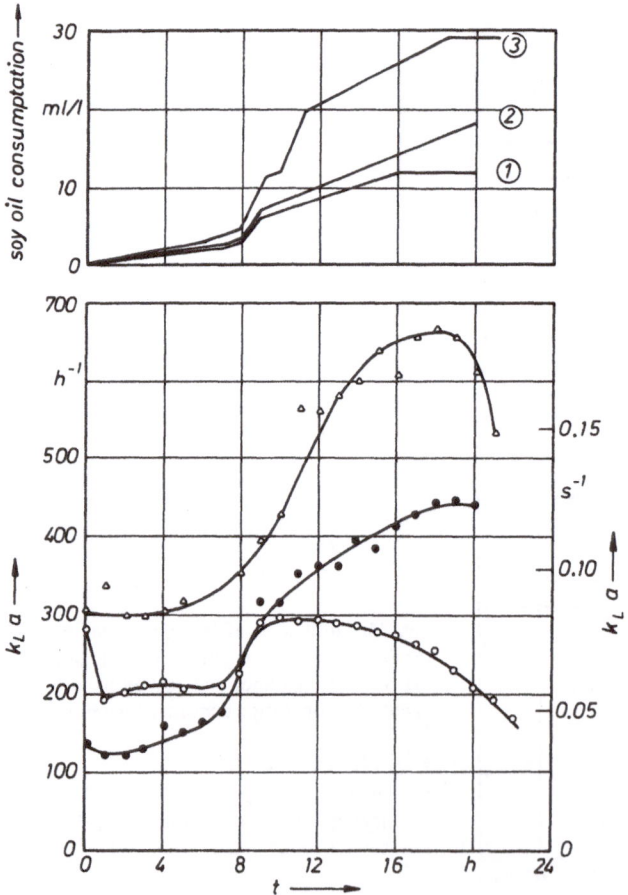

Fig. 86. Volumetric mass transfer coefficients, $k_L a$'s, as a function of the cultivation time, t, during the cultivation of *H. polymorpha* in Reactors C (single-stage), D-40 (three-stage, h = 40 cm) and D-20 (three-stage, h = 20 cm) employing substrate ethanol and soy oil as antifoam agent. (For operation conditions see Fig. 81) [87, 88];

o 1   single-stage tower loop
● 2   three-stage tower loop, h = 40 cm
△ 3   three-stage tower loop, h = 20 cm

top: soy oil added to the system (ml l⁻¹)

tower lowest, as expected. The strong increase of $k_L a$ in the presence of large amounts of antifoam agent (curve 3 in Fig. 86) was also found by determining the $k_L a$ of cultivation media in a twin bubble column employing different amounts of antifoam agents [217]. The oxygen transfer rates increase with increasing time. For $t < 12$ h the OTR in the single-stage tower was higher than in the three-stage one with $h = 40$ cm. Again no explanation can be given for this unusual behaviour. For $t > 12$ h, the OTR in the single-stage tower with $h = 20$ cm is highest and in single-stage tower lowest, as expected. The maximum OTR values are 2.50 g l^{-1} h^{-1} in the three-stage tower with $h = 20$ cm and 1.60 g l^{-1} h^{-1} in the single-stage tower. The maximum cell productivities correspond to these data: $Pr = 2.93$ g l^{-1} h^{-1} (three-stage tower, $h = 20$ cm) and $Pr = 1.69$ g l^{-1} h^{-1} (single-stage tower).

7.3 Comparison of Different Systems

It is only possible to compare different reactors, if the same biological system (strain, medium, growth conditions) are employed in them. In the following, this comparison will be restricted to such parameters as
— construction and operation conditions
— influence of medium composition.

7.3.1 Influence of Construction Parameters and Operation Conditions

With increasing gas flow rate, $\dot{V}_G$, the specific power input, P_L/V_L, is enlarged.

Figures 87a, 87b and 87c show how $k_L a$ increases with increasing specific power input, P_L/V_L in a counter current multistage tower reactor by employing different model media and three different aerators. Such relationships in tower reactors (without a loop) are expected based on Eq. (117):

$$k_L a = \text{const} \left(\frac{P_L}{V_L} \right)^m, \tag{125}$$

The exponent, m, is influenced by different parameters [1].

However, in tower loop reactors $\dot{V}_G$ also influences several other parameters: medium recycling ratio in air lift reactors, longitudinal medium dispersion, cell floration, cell sedimentation/flocculation, etc. Therefore the relationship between cell mass productivity and gas flow rate is complex. E.g. because of the air lift effect, OTR and Pr pass maximums with increasing $\dot{V}_G$ (Fig. 14). At a given gas flow rate, $\dot{V}_G$, the relative mean gas hold up, E_G, depends on the operation conditions. In the homogeneous flow range the effective linear gas velocity, $w_G = \dfrac{\dot{V}_G}{QE_G} = \dfrac{w_{SG}}{E_G}$, in the tower is given in concurrent upstream flow by:

$$\frac{w_{SG}}{E_G} = \frac{w_{SL}}{1 - E_G} + w_{BR} \tag{126}$$

in concurrent downstream flow:

$$-\frac{w_{SG}}{E_G} = -\frac{w_{SG}}{1 - E_G} + w_{BR}, \quad \text{with} \quad \left| \frac{w_{SL}}{1 - E_G} \right| > |w_{BR}| \tag{127}$$

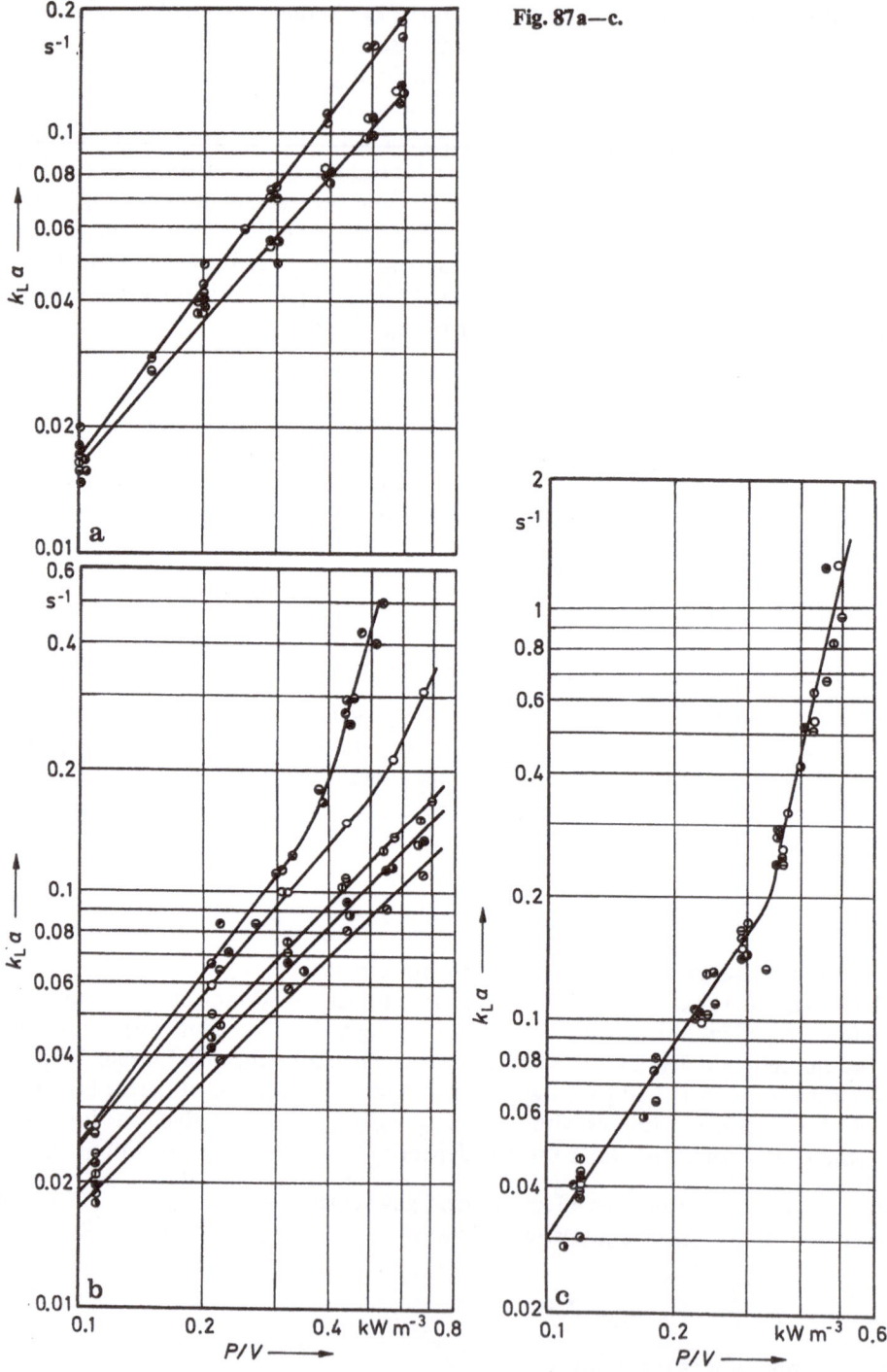

Fig. 87a—c.

and in countercurrent flow:

$$\frac{w_{SG}}{E_G} = -\frac{w_{SG}}{1 - E_G} + w_{BR}, \quad \text{with} \quad \left|\frac{w_{SL}}{1 - E_G}\right| < |w_{BR}| \tag{128}$$

In these equations, w_{BR} is the relative bubble swarm velocity with regard to the liquid and depends on the bubble size.

w_{SL} is enlarged by increasing w_{SG} in air lift reactors, e.g., according to Weiland [148]:

$$w_{SL} = 1.15 w_{SG}^{0.414} \tag{129}$$

where w_{SL} and w_{SG} (m s^{-1}) are the superficial liquid and gas velocities in the tower. With increasing w_{SG}, $\dot{V}_R$ increases too. In tall reactors the exponent in Eq. (129) is probably larger which results in still higher $\dot{V}_R$'s. In these all tower loop reactors $w_L = w_{SL}/(1 - E_G)$ and can have very high values. Under given operation conditions (w_{BR} and w_{SG} are fixed), but with increasing w_L Eq. (126) can only be fulfilled, if E_G diminishes and at very high w_L approaches zero. Thus E_G, a and $k_L a$ will be considerably reduced.

A concurrent downstream flow can only be maintained according to Eq. (127) if the liquid velocity is higher than the relative bubble swarm velocity, w_{BR}. All bubbles will rise at a given w_L against the liquid flow which have higher bubble rise velocities than w_L. If a complete separation of bubbles is desired the downstream liquid flow rate, w_L, should be low (low $\dot{V}_R$ or horizontal flow section). According to Eq. (128) a countercurrent flow can be maintained if $w_{BR} > w_L$. Flooding occurs if $w_{BR} < w_L$.

E_G is the highest in concurrent downstream flow, the smallest in concurrent upstream flow and intermediate in countercurrent flow as can be seen from Eqs. (126—128).

The difference in E_G between the concurrent upstream flow (Reactor A) and countercurrent flow (Reactor C) depends on the ratio of the effective liquid velocity,

$$w_L = \frac{w_{SL}}{1 - E_G}, \quad \text{and } w_{BR}.\ \text{In Reactor C at } \dot{V}_R = 200\ \text{l h}^{-1},\ w_L \simeq 0.22\ \text{cm s}^{-1}\ \text{and}$$

$w_{BR} \simeq 22$ cm s^{-1}. E_G is only slightly influenced by the direction of the liquid

◀ **Fig. 87a.** Volumetric mass transfer coefficient, $k_L a$, in a six stage tower reactor 20 cm in diameter, h = 30 cm, perforated plate trays 3 mm in hole diameter with different media as a function of specific power input, P_L/V_L [87];
● 1% methanol solution
○ 1% ethanol solution
◑ 1% n-propanol solution
◉ 0.5% n-propanol solution
◔ 1% glucose solution
◕ *H. polymorpha* nutrient salt solution (Table 7)
◒ *H. polymorpha* nutrient salt solution, 1% methanol
◓ *H. polymorpha* nutrient salt solution, 1% ethanol;
b. Volumetric mass transfer coefficient, $k_L a$, in a six stage tower reactor h = 30 cm, 20 cm in diameter, perforated plate trays 1 mm in hole diameter with different media as a function of specific power input, P_L/V_L [87]. (For symbols see Fig. 87a);
c. Volumetric mass transfer coefficient, $k_L a$, in a six-stage tower reactor h = 30 cm, 20 cm in diameter, perforated plate trays 0.5 mm in hole diameter with different media as a function of specific power input, P_L/V_L [87]. (For symbols see Fig. 87a)

flow. Thus, the difference between Reactors A and C when employing the same biological system (*H. polymorpha*) are mainly due to different *aerators*. In Reactor A, a porous plate (17.5 µm) and in Reactor C, a perforated plate (0.5 cm in hole diameter) were employed.

Since *H. polymorpha* medium has considerable bubble coalescence suppressing character, the differences in d_S were preserved in the tower. In spite of the larger bubbling layer height of the Reactor A, d_S was smaller in Reactor A, and as a consequence, OTR, k_La and Pr were larger in Reactor A than in C (Table 9).

A perforated plate 0.5 mm in hole diameter with 0.35 % relative free surface area is indeed quite an inefficient aerator in comparison to the porous plate (see e.g. Fig. 7), but it is still much better than a perforated plate 3.0 mm in hole diameter with 6.52 % relative free surface area.

In Reactor A a highly efficient porous plate aerator was employed. In the ten-stage tower loop (Reactor B) nine perforated plates 3.0 mm in hole diameter were additionally installed as stage separating trays. Since a liquid free gas layer was formed below each of the trays, they acted as gas distributors, i.e. the bubbles were newly formed in each of the stages by these very inefficient aerators. The result is a considerable deterioration of reactor performance in comparison with the single-stage tower loop (Reactor A) (Table 10). One can recognize from this table that the difference between Reactors A and B is much less in batch operation than in continuous operation. Considerable is the difference between reactor performances in batch and continuous operations: μ_m seems to be higher and $Y_{X/S}$ lower in batch than in continuous reactor in contrast to stirred tank reactors in which μ_m, $Y_{X/S}$ as well as Pr are higher in the continuous reactor than in the batch one for the same biological system (Table 11). On the other hand, one can recognize from Tables 10 and 11 that μ_m evaluated in batch reactors cannot be employed for continuous reactors and μ_m measured in stirred tank reactors cannot be used for tower loop reactors.

In contrast to the poor performance of Reactor B with regard to Reactor A, Reactor D exhibits a better performance than Reactor C. In Reactor C, a perforated plate 0.5 mm in hole diameter with 0.35 % relative free surface area is used as a gas distributor. Additionally in Reactor D the same perforated plates were used as stage separating trays. Since below every tray a liquid free gas layer existed, each of the trays acted as a gas distributor. By eliminating the foam with Desmophen the efficiency of these trays was preserved. Thus, efficiencies of the primary and secondary gas dispersions were the same. It is well known from model medium

Table 9. Comparison of the performances of Reactors A and C employing *H. polymorpha* on ethanol substrate. $Y_{X/O}$, k_La, OTR and P_r given for oxygen transfer limited growth range

	Reactor A	Reactor C
	Porous plate 17.5 µm pore diameter	Perforated plate 0.5 mm hole diameter
μ_m (h^{-1})	0.26	0.25
$Y_{X/O}$ (−)	0.53	0.55
OTR (g l^{-1} h^{-1})	4.7	1.28
k_La (h^{-1})	1200	250
Pr (g l^{-1} h^{-1})	2.5	0.71

Table 10. Comparison of the performances of Reactor A (single-stage, with porous plate aerator) and Reactor B (ten-stage, with porous plate aerator and 9 trays, perforated plates 3.0 mm in hole diameter) employing *E. coli* and antifoam agent (Desmophen) in batch and continuous operation [305)]

Operation	Batch, $\dot{V}_R = 340\,l\,h^{-1}$		Continuous, $\dot{V}_R \doteq 1400\,l\,h^{-1}$	
Stages	Single	Ten	Single[b]	Ten[b]
Gas distribution	Porous plate	Perforated plate trays	Porous plate	Perforated plate trays
μ_m (h^{-1})	0.54	0.52	0.36–0.44	0.36
$Y_{X/S}^c$ (—)	0.80	0.75	1.16	0.53
RQ (—)	0.88	0.93	0.93	0.86
Pr, Pr[a] $(g\,l^{-1}\,h^{-1})$	1.63	1.66	2.64	1.15

[a] $Pr = DX$ in continuous culture;

$Pr = \dfrac{\Delta X}{\Delta t}$ in batch culture (not considering the lag phase);

[b] $\mu = D$

[c] formal value calculated with regard to glucose

Table 11. Comparison of batch and continuous cultivations in stirred tank reactor by employing *E. coli* and antifoam agent (Desmophen) [305)]

Cultivation	Batch	Continuous
μ_m (h^{-1})	0.90	1.10
$Y_{X/S}$ (—)	0.70	0.90
RQ (—)	0.87	0.85
Pr, Pr[a]	1.25	2.0

[a] $Pr = DX$ in continuous; $Pr = \dfrac{\Delta X}{\Delta t}$ (lag phase is excluded)

measurements, that the stage separating trays strongly influence $k_L a$, as can be seen from Fig. 88. Thus the gas dispersion efficiency of the trays is an important factor for multistage tower reactors. One can recognize from Table 12 that the three-stage tower loop (Reactor D) has a higher performance than the single stage one (Reactor C), if antifoam agents are employed. (In the absence of antifoam agents cell microflotation deteriorates the performance of Reactor D, with regard to cell productivity.)

One can also recognize from Table 12 that the height of the bubbling layer also influences the multistage tower loop reactor performance. This influence has been well represented by means of model media (Fig. 89). As expected $k_L a$, OTR and Pr increase with diminishing height of the bubbling layer, h. This phenomenon is due to the higher k_L-value which prevails immediately after bubble formation and the smaller d_S caused by the slight bubble coalescence in this short layer. The liquid recirculation ratio, influences the performance of tower loop reactors in different ways. Increasing γ reduces E_G, a and $k_L a$, if OTR is dominating (Fig. 64). It has an optimum value, if substrate limitation prevails (Chapter 7.1.1/7.1.2) and it can

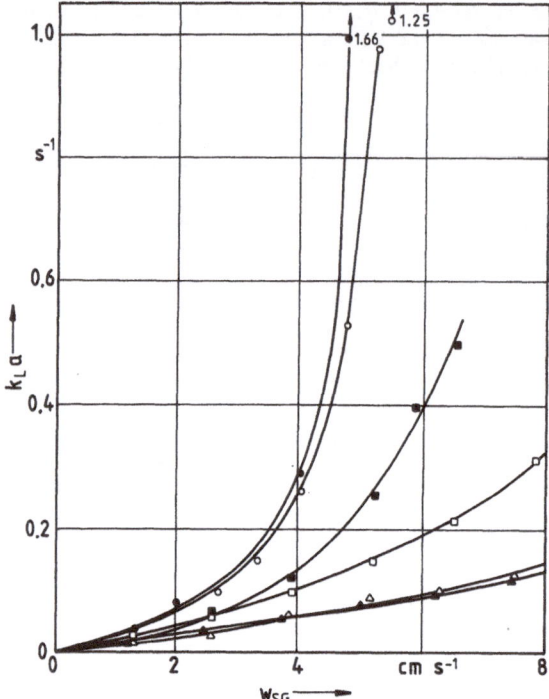

Fig. 88. Volumetric mass transfer coefficient, k_La, in a six-stage tower reactor 20 cm in diameter, as a function of the superficial gas velocity, w_{SG} for different trays. 1% ethanol and nutrient salt solutions [87] perforated plate hole diameter, d_H, number of holes, N_H, relative free cross section Q_H

d_H (mm)	N_G (—)	Q_H (%)	1% ethanol	nutrient salt sol.
3	214	3.82	△	▲
1	248	0.62	□	■
0.5	558	0.35	○	●

influence the longitudinal medium dispersion, especially the longitudinal cell mass concentration profiles. The latter effect is especially important, if cell sedimentation due to cell flocculation is considerable. In this case, with increasing γ, the tower loop reactor performance improves (Table 13). With decreasing γ, substrate conversion, U_s, and yield coefficient, $Y_{X/S}$ diminish in single-stage tower loop reactors. Also the cell productivity, $Pr = DX$, diminishes from $2.64 \, g \, l^{-1} \, h^{-1}$ ($\gamma = 70$) to $1.48 \, g \, l^{-1} \, h^{-1}$ ($\gamma = 0$). The same holds true for ten-stage tower loop reactors (Table 13). The cell productivity, $Pr = DX$, also diminishes from $1.15 \, g \, l^{-1} \, h^{-1}$ ($\gamma = 13.8$) to $1.08 \, g \, l^{-1} \, h^{-1}$ ($\gamma = 0.95$) if the recirculation ratio, γ, is reduced.

7.3.2 Influence of Medium Composition

The influence of substrate type on the biological parameters is well documented (Tables 4 and 5). In the present case μ_m is much higher for glucose than for ethanol or methanol. Also the substrate and oxygen yield coefficients as well as

Table 12. Comparison of single- and three-stage counter current tower loops (Reactors C and D) by employing *H. polymorpha* in antifoam agents [87, 88, 89]

Antifoam agent	AA Absence		Desmophen 3600			Soy oil		
	Single-stage	Three-stage	Single-stage	Three-stage		Single-stage	Three-stage	
H (cm)	(160)[a]	40[c]	(160)[a]	40	20	(160)[a]	40	20
μ_m (h^{-1})	0.25	0.17	0.21	0.21	0.21	0.24/0.26[b]	0.11/0.26[b]	0.09/0.27[b]
k_La^d (h^{-1})	250	800	160	220	250	290	440	650
(s^{-1})	0.07	0.22	0.04	0.06	0.07	0.08	0.12	0.18
$Y_{x/o}\left[\dfrac{\text{g biomass}}{\text{g O}_2}\right]^e$	0.55	0.36	0.69	0.58	0.74	1.06	1.16	1.17
Pr [g l^{-1} h^{-1}]^e	0.71	0.80	0.68	0.75	1.09	1.69	2.43	2.93
Q_{O_2m} [g l^{-1} h^{-1}]	1.28	2.20	0.98	1.30	1.48	1.60	2.10	2.50
spec. energy requirement $\left[\dfrac{\text{kW h}}{\text{kg O}_2}\right]$	0.20	0.14	0.27	0.23	0.20	0.16	0.14	0.12
spec. energy requirement $\left[\dfrac{\text{kW h}}{\text{kg biomass}}\right]$	0.37[f]	0.38[f]	0.38	0.40	0.28	0.15	0.12	0.10

[a] Mean bubbling layer height;
[b] First exp. growth phase/sec. exp. growth phase;
[c] Large error due to foam formation;
[d] At maximum oxygen transfer rate;
[e] In the oxygen transfer limited linear growth phase;
[f] Without the energy requirement of the mechanical foam destroyer

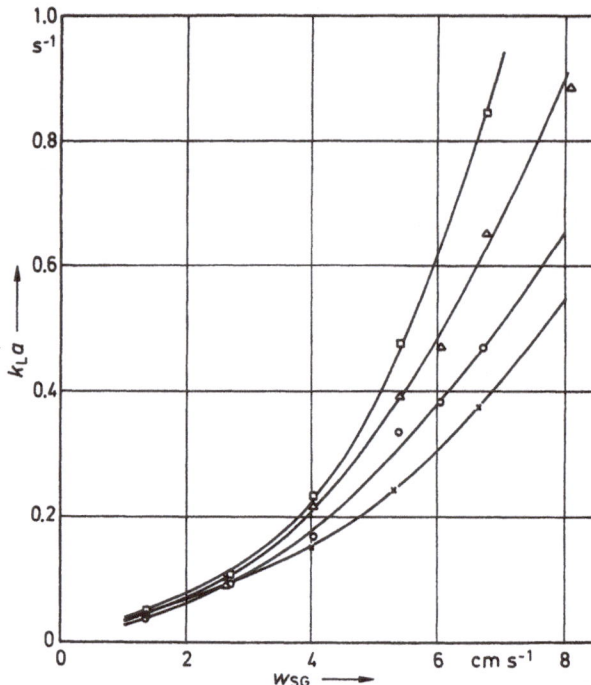

Fig. 89. Volumetric mass transfer coefficient, $k_L a$, in a six stage tower reactor 20 cm in diameter as a function of the superficial gas velocity, w_{SG}, for different heights of the bubbling layer, h. Perforated plate tray 0.5 mm in hole diameter, d_H, distilled water [87];

□ h = 10 cm   ○ h = 30 cm
△ h = 20 cm   × h = 50 cm

Table 13. Influence of the recirculation rate on substrate conversion, U_s, and yield coefficient, $Y_{X/S}$, in single- and ten-stage tower loop reactors by employing *E. coli* and antifoam agent [305]

Single-stage			Ten-stage		
γ (—)	U_s (%)	$Y_{X/S}$ (—)	γ (—)	U_s (%)	$Y_{X/S}$ (—)
70	ca. 100	1.16	14	98.2	0.63
3	95.7	1.15	6	76.8	0.47
0	88.5	0.80	1	49.4	0.48

respiratory quotients are very different (Table 14). However, these coefficients were not constant during the cultivation, e.g. in substrate limited growth oxygen yield coefficients decreased with diminishing OTR.

The substrate, cell, and oxygen concentrations have considerable influence on the performance of tower loop reactors. In Fig. 90, the influence of cell mass concentration on $k_L a$ is shown during the cultivation of *H. polymorpha*. Except for the range X < 8 g l, in which other factors dominate, $k_L a$ gradually diminishes with increasing X, if ethanol substrate is employed. When using glucose, $k_L a$ increases with increasing X, due to the increasing ethanol concentration produced by the cells.

The strong ethanol effects on $k_L a$ and a are shown in Fig. 91. The initial time range (t < 9) should be disregarded here because of the strong environmental cell control in this range. One can clearly recognize from Fig. 91 that with diminishing ethanol concentration $k_L a$ and a are reduced, and at increasing concentration they are enlarged.

Table 14. Comparison of single-stage concurrent tower loop reactors by employing *H. polymorpha*, ethanol and glucose as substrate. Biological parameters of the investigated systems [16, 153, 282]

Substrate	Growth phase	μ_m	$Y_{X/S}$ $\dfrac{g\,X}{g\,substrate}$	$Y_{X/O}$ $\dfrac{g\,X}{g\,O_2}$	R.Q. $\dfrac{mol\,CO_2}{mol\,O_2}$
Ethanol	Exponential	0.26	0.25–0.55	0.32–0.53	0.48
	O_2-Limited		0.50–0.75	0.25–0.53	0.48
	S-Limited		0.55–0.75	0.32–0.56	0.48
Glucose	Exponential	0.58	0.47	1.16–1.36	1.0
	O_2-Limited		0.46	0.65–1.24	1.0
	S-Limited		0.46	0.98–1.21	1.0

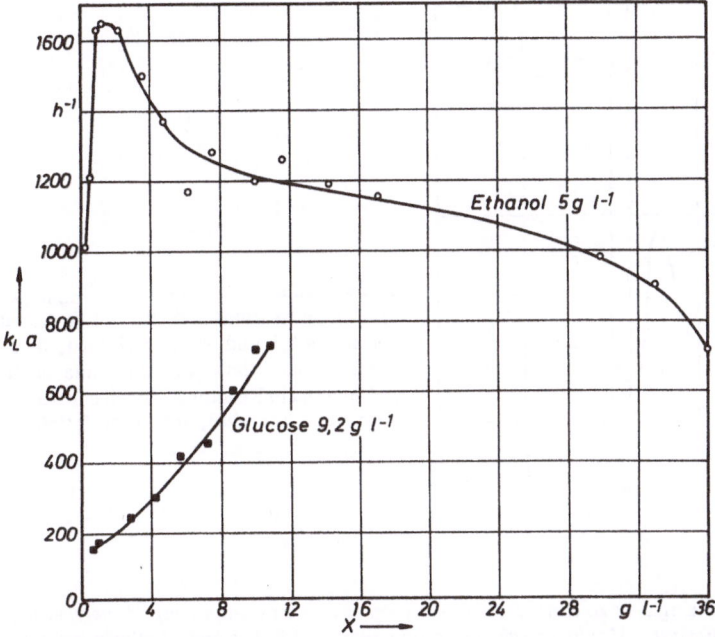

Fig. 90. Volumetric mass transfer coefficient, $k_L a$, as a function of the cell mass concentration, X, during the cultivation of *H. polymorpha* in Reactor A employing substrates glucose and ethanol (extended culture) [16, 154];
o ethanol $S = 5 \ g\,l^{-1}$
■ glucose $S = 9.2 \ g\,l^{-1}$

The absence of different essential medium components, of course, stop the growth. However, the cells react differently on their absence. Fig. 92 shows that in the absence of ethanol, cell growth stagnates, OTR drops to zero, and dissolved oxygen saturation approaches unity. However, the temporary lack of substrate does not influence μ_m (Fig. 93). The same is true for glucose substrate (Figs. 94 and 95). However, a 2.5 min oxygen lack in the medium reduced μ_m from $0.26 \ h^{-1}$ to $0.13 \ h^{-1}$ (Fig. 96).

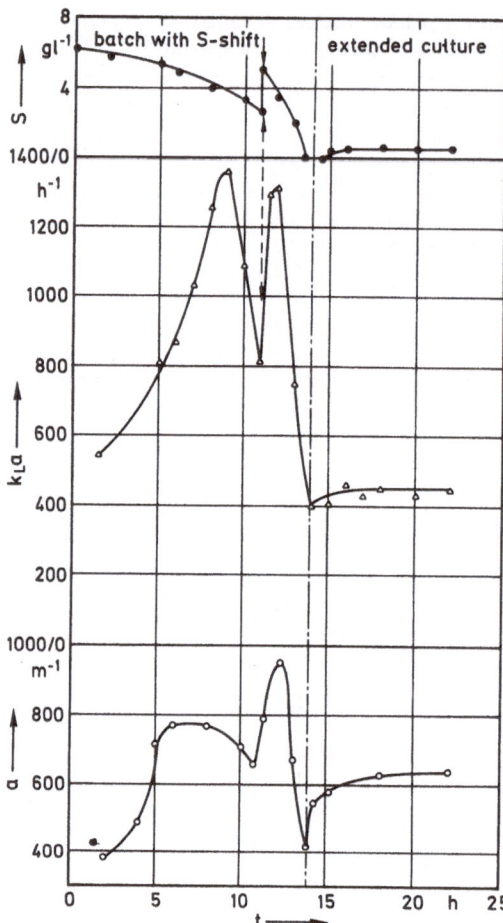

Fig. 91. Ethanol concentration, C_{EtOH}, volumetric mass transfer coefficient, $k_L a$, and specific gas/liquid interfacial area, a, as a function of the cultivation time of *H. polymorpha* in Reactor A. Batch cultivation with substrate shift, 0.55 vvm [16, 155];
● C_{EtOH}; △ $k_L a$; ○ a

Table 15. Comparison of single-stage concurrent tower loop reactors by employing *H. polymorpha*, ethanol and glucose as substrate. *Maximum values* of productivity, OTR, $k_L a$ and a in the investigated systems [16, 153, 282]

Substrate	Growth phase	Pr g l^{-1} h^{-1}	Q_{O_2} g l^{-1} h^{-1}	$k_L a$ h^{-1}	a m^{-1}
Ethanol	O_2-Limited	2.5	4.7	1200	1400
	S-Limited	0.95	1.7	300	400
Glucose	O_2-Limited	3.3	2.8[a]	540[a]	650[a]
	S-Limited	1.7	1.4	300	300

[a] with ethanol production

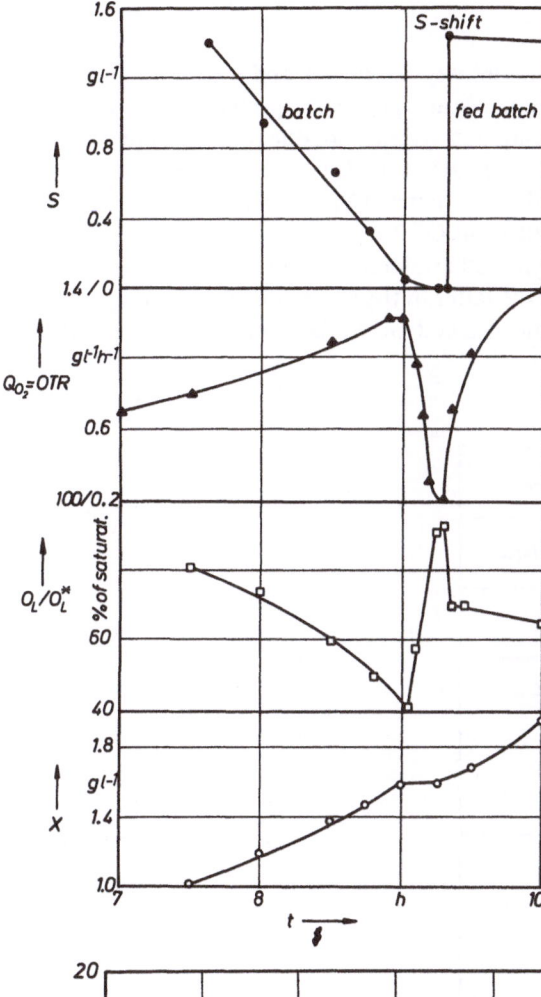

Fig. 92. Substrate ethanol concentration, S, OTR, DOS and (dry) cell mass concentration, X, in Reactor A as a function of the cultivation time, t, of *H. poly-morpha*. Variation of the substrate concentration [16, 155];

● C_{EtOH}, ▲ OTR, □ O_L/O_L^*
○ (dry) cell mass concentration, X

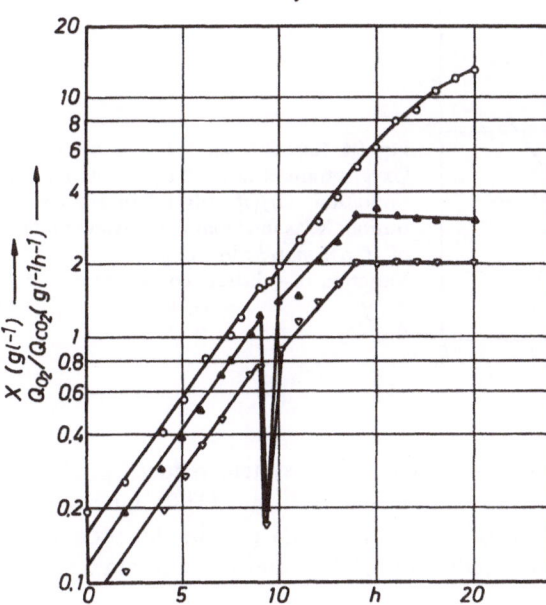

Fig. 93. Cultivation of *H. polymorpha* on substrate ethanol in Reactor A. Influence of a short time lack of substrate. Aeration rate 0.55 vvm [16, 155];

○ (dry) biomass concentration, X
▲ oxygen uptake rate
▽ CO_2 production rate

In Table 15, maximum values of productivity, OTR, k_La and a are compiled for *H. polymorpha* cultivations on substrates ethanol and glucose respectively. The differences between OTR, k_La and a in O_2- and S-limited systems are due to the different ethanol concentrations. When using ethanol substrate, this difference is larger, because of the higher ethanol concentration in these systems. The differences in Pr in O_2- and S-limited systems are caused by the different OTR's.

The differences in Pr when employing ethanol and glucose substrate, respectively are mainly due to the higher $Y_{X/O}$ for the latter in the O_2-limited growth range. In the S-limited range the higher Pr in the glucose system can be explained by the higher μ value.

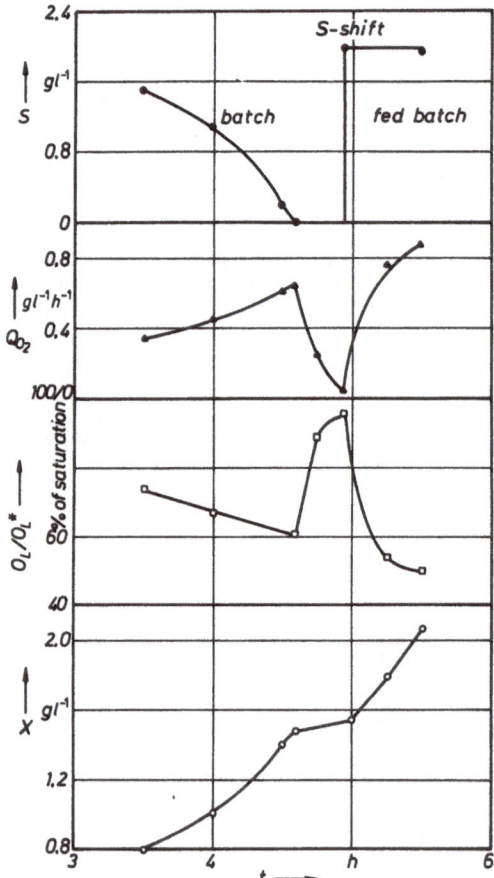

Fig. 94. Substrate glucose concentration, S. Oxygen transfer rate, OTR, dissolved oxygen saturation, O_L/O_L^*, (dry) biomass concentration, X, as functions of cultivation time, t, of *H. polymorpha* in Reactor A
Variation of substrate concentration [16, 155];
● S   □ O_L/O_L^*
▲ Q_{O_2} ○ X

The addition of antifoam agent (Desmophen) to the system reduced μ_m (e.g. in Table 12 from 0.25 to 0.21 h^{-1}). It also diminishes k_La and OTR, but increases $Y_{X/O}$ (see single-stage in the absence of antifoam agents and in the presence of Desmophen in Table 12). The strong reduction of k_La due to antifoam agents is also well documented in Ref. [218].

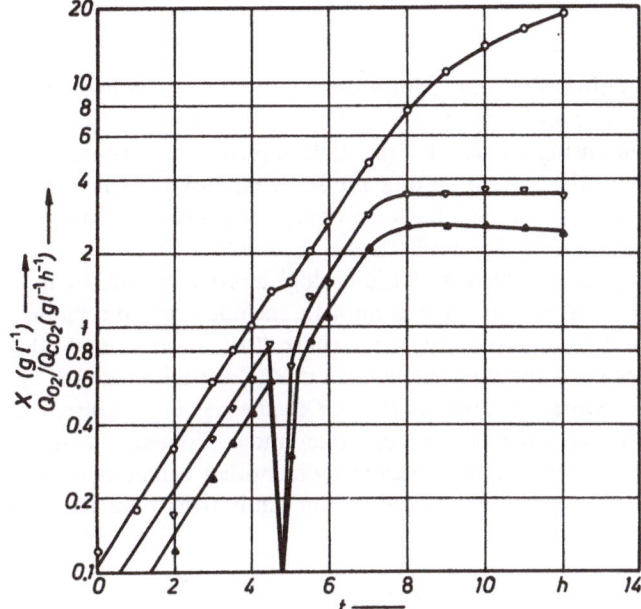

Fig. 95. Cultivation of *H. polymorpha* in Reactor A employing substrate glucose. Influence of the short time lack of substrate. Aeration rate: 0.55 vvm [16, 155]. (For symbols see Fig. 93)

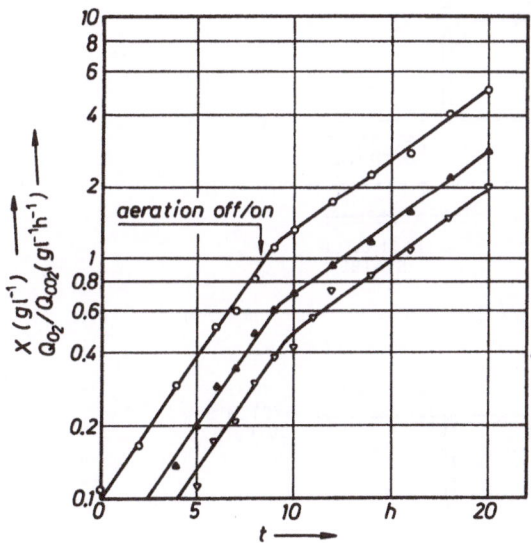

Fig. 96. Extended cultivation of *H. poly-morpha* in Reactor A employing substrate ethanol. Influence of the short term lack of dissolved oxygen caused by interruption of the aeration. $S = 5$ g l^{-1}, 0.55 vvm [16, 155]. (For symbols see Fig. 93)

8 Control Strategies

Neither batch nor fed batch cultures or start procedures of continuous cultures can sufficiently be controlled by classical techniques. Different methods are available to solve these problems: When using an *observer* (for deterministic systems) or *filter* (for noisy systems) to estimate the system state a model is employed, which runs parallel to the equipment and the differences between the system and model outputs are used to improve the estimation of the system states (Fig. 97) [311, 312].

Because of temporarily varying parameters of the biological systems and the non linearities of the growth processes, which often cannot be included into the model, the *adaptive controller (model reference adaptive controller)* insures that the differences between the outputs of the equipment and model are minimized (Fig. 98) [309, 310]. Another adaptive control strategy is the *self tuning regulator*. Here, the classical feed back scheme with reference inputs, controller and equipment is entirely maintained; only the control parameters are adjusted by identified values of system parameters. Identification is performed by means of an equipment model. Again

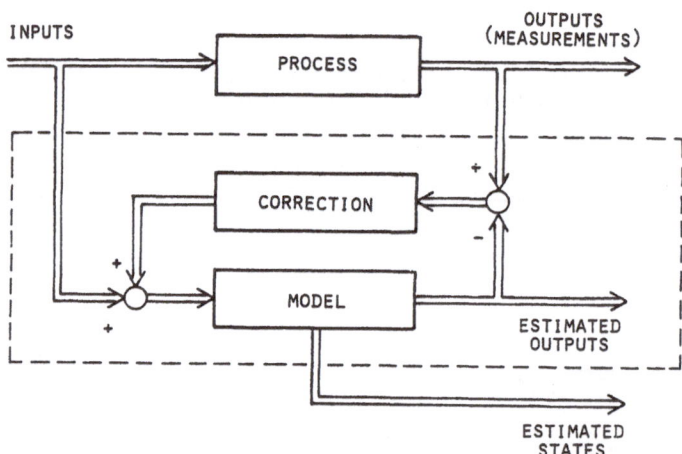

Fig. 97. Structure of a state estimator [309, 310]

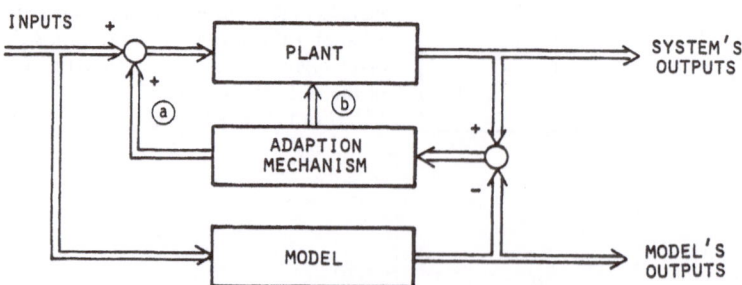

Fig. 98. Model reference adaptive control [309, 310]; ⓐ signal synthesis adaptation; ⓑ parameter adaption

the outputs of model and equipment are compared and their difference is used to improve estimations for the unknown system parameters (Fig. 99) [309, 310].

The models employed in these advanced control techniques should be sufficiently exact, but not too extensive, because they are to be stored in a process computer.

It is a difficult task to find the optimal reduction of the equipment model. Extensive models developed by Luttmann [181, 269–271, 308, 315, 316] are able to correctly simulate the biochemical processes in tower loop reactors; e.g. (Fig. 100), however, they are too large for process computers. Therefore Munack reduced these models and used them to compare the optimal control with estimated parameters the adaptive control with identifications and the optimal control with known parameters according to the open loop feed back optimal control (Fig. 101) [309].

His investigations show that it is possible to control tower loop reactors by means of adaptive control and parameter identification by employing a simplified distributed parameter model [309, 310]. This will be treated in a separate paper [313].

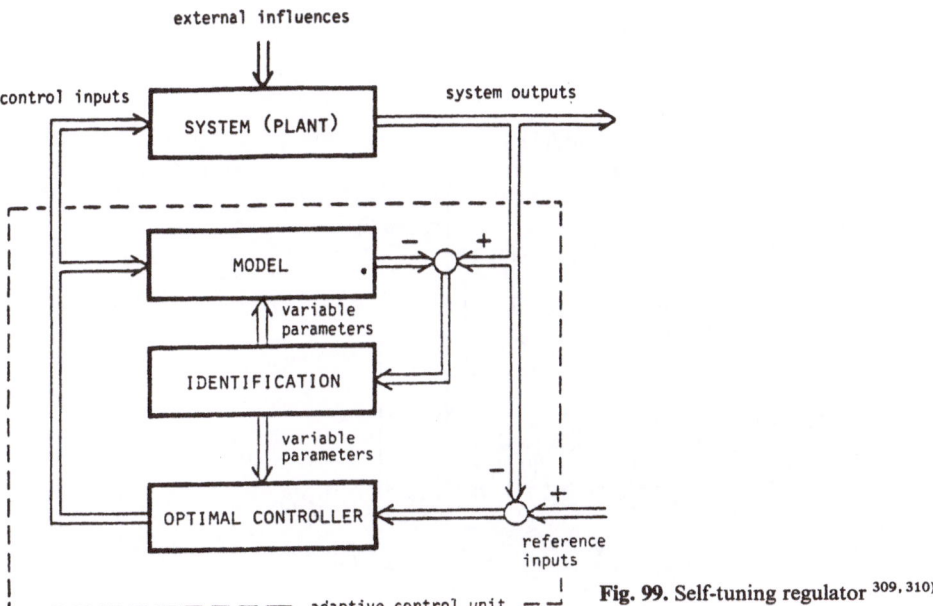

Fig. 99. Self-tuning regulator [309, 310]

9 Outlook

Because of the complexity of biological systems, time and space dependence of several parameters and their partly unknown interrelations, it is a difficult task to characterize and predict the performance of tower loop reactors. It is necessary to develop and improve on-line methods to measure biological activity of cells, chemical and physico-chemical properties of cultivation media, physical properties of the two-phase flow and evaluate their interrelations. Furthermore, we need to develop better models which include not only cells with invariable properties, but

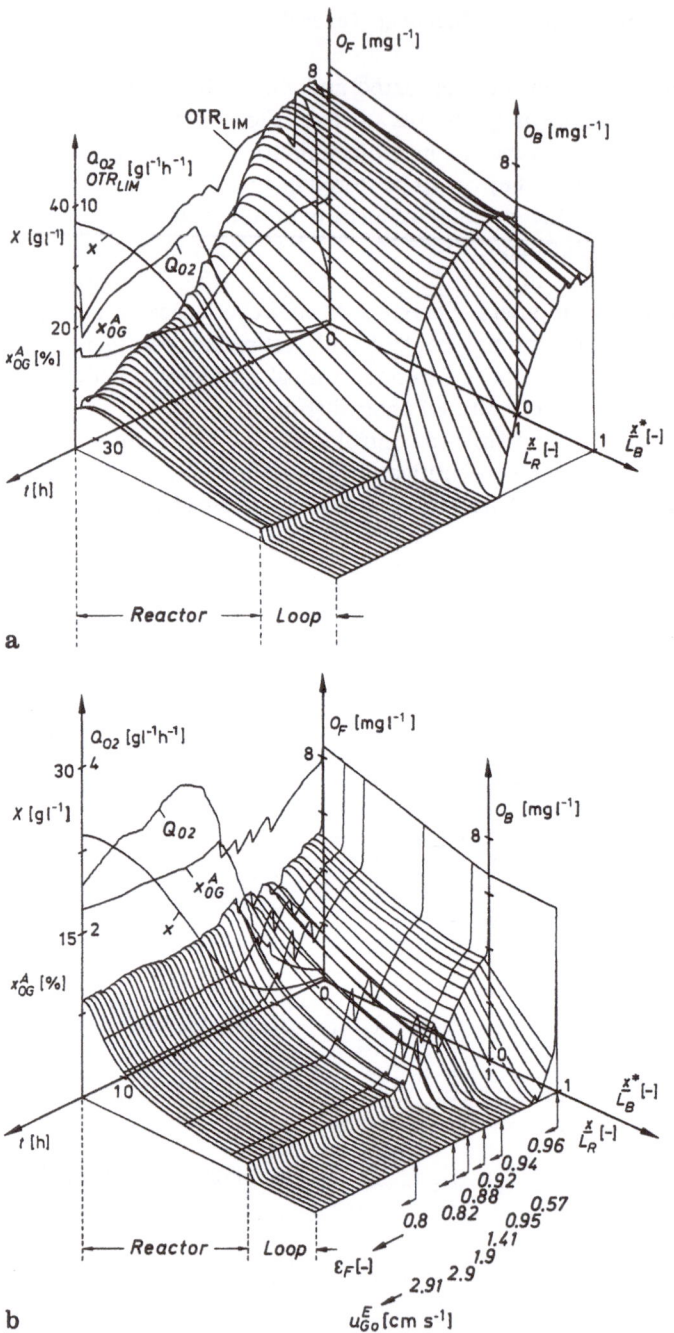

Fig 100 a and b. Comparison of the courses of extended culture cultivations of *H. polymorpha* in Reactor A. I. on substrate ethanol (coalescence suppressing medium) and II. on substrate glucose (coalescence neutral medium). Spacial and time variations of dissolved oxygen concentration in tower, O_F and loop, O_B, as well as OTR_{LIM} (upper limit), OTR (actual) oxygen concentration in the gas phase, x^A_{OG}, at the exit, cell mass concentration, X, as function of the cultivation time, t. The variation of mean relative liquid hold up, ε_F, and superficial gas velocity at the entrance, u^E_{Go} with the time.

ⓐ Substrate ethanol, S = 5 g l^{-1};
ⓑ Substrate glucose, S = 2 g l^{-1};
These curves were simulated by means of the generalized model and identified parameters [181]

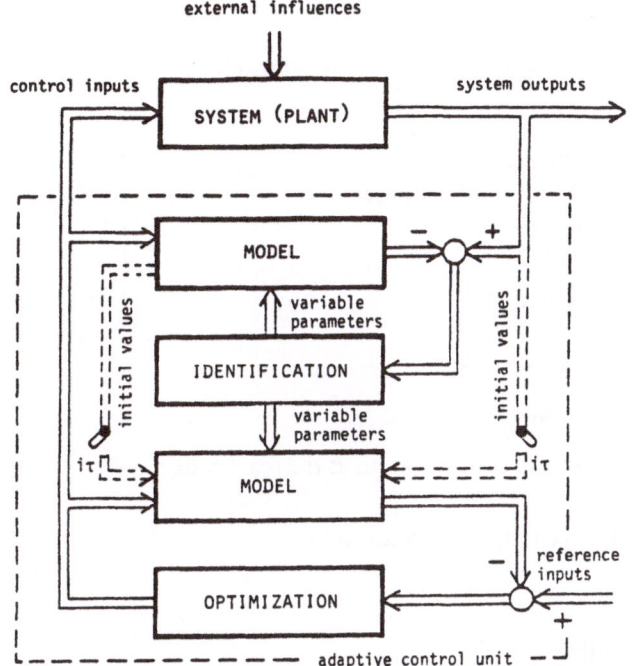

Fig. 101. Open loop feedback optimal control (OLFO) [309, 310]

also their adaptation to their environment under different operation conditions (substrate, oxygen transfer limited growth ranges) and such phenomena as micro-flotation as well as cell sedimentation and flocculation.

It is expected that by means of this additional information and improved models it will be possible to simulate the complete process in pilot and production plants based on the bench scale reactor measurements and to control it by one of the advanced techniques.

A good beginning was made by Luttmann [181], who simulated an operating pilot plant reactor by means of the bench scale masurements of H. Buchholz [16] and Lehmann et al. [26] as well as by the information provided by the pilo-plant operating company, assuming different economical conditions.

Further developments along this line are necessary to realize these goals.

10 Acknowledgement

The author acknowledges the financial support of the Ministry of Research and Technology of the Federal Republic of Germany, Bonn, and the German Research Foundation (DFG), Bonn, and thanks Dr. I. Adler, Dr. H. Buchholz, Dr. R. Buchholz, Dr. R. Luttmann, Dr. J. Voigt, Dr. W. Zakrzewski and Mr. J. Lippert for their excellent cooperation as well as Prof. Dr. H. Sahm, Institute of Bio-technology, KFA Jülich, Prof. Dr. M. Thoma and Dr. A. Munack, Institute for Control and Automation of the University of Hannover for their support.

11 Nomenclature

L (length), M (mass), T (time), K (temperature), — (dimensionless)

A	area	L^2
A	gas/liquid interfacial area	L^2
a'	geometrical gas/liquid interfacial area	L^2
$a = A/V_L$	specific interfacial area	L^{-1}
a	a calculated by Eq. (35)	L^{-1}
$a^* = \dfrac{A}{V}$	specific interfacial area	L^{-1}
$Bo = \dfrac{u_F H}{D_F}$	Bodenstein-number in the tower	—
BSA	Bovine Serum Albumine	
$C_s = \dfrac{S_e}{S_0}$	dimensionless substrate concentration at the reactor exit	—
$C_x = \dfrac{X}{X + Y_{X/S} S_0}$	dimensionless cell mass concentration	—
c	concentration	ML^{-3}
D	dilution rate	T^{-1}
$Da = \mu_m \tau,$	Damköhler number	—
D_M	diffusivity	$L^2 T^{-1}$
D_F	longitudinal liquid dispersion coefficient in tower	$L^2 T^{-1}$
D_G	longitudinal gas dispersion coefficient in tower	$L^2 T^{-1}$
$\bar{d}_B$	mean bubble diameter Eq. (32)	L
d_S	Sauter bubble diameter Eq. (33)	L
d_e	dynamical equilibrium bubble diameter	L
d_p	primary bubble diameter	L
E_{O_2}	aeration efficiency	$T^2 L^{-2}$
E_G	mean relative gas hold-up, Eq. (31)	—
E_F	mean relative liquid hold-up	—
E	energy dissipation rate (local power input)	$ML^2 T^{-3}$
$E_1(n)$	one dimensional frequency power spectrum	$L^3 T^{-3}$
$E_1(k)$	one dimensional wave number power spectrum	$L^2 T^{-1}$
F	liquid feed rate	$L^3 T^{-1}$
g	acceleration of gravity	LT^{-2}
H	height of the aerated layer	L
He	Henry coefficient	$T^{-2} L^2$
H_L	height of the bubble free liquid layer	L
h	height of the aerated layer in multistage tower	L
I	ion strength Eq. (6)	ML^{-3}
$I = u'/\bar{U}$	relative turbulence intensity	—
$K = \dfrac{K_s}{S_0 + X_0/Y_{X/S}}$	dimensionless substrate saturation constant	—
K_0	oxygen saturation constant	ML^{-3}
K_s	substrate saturation constant	ML^{-3}
K_{ST}	coalescence constant, Eq. (97)	—

$k = 2\pi n/\bar{U}$	wave number	L^{-1}
k_L	mass transfer coefficient	LT^{-1}
$k_L a$	volumetric mass transfer coefficient	T^{-1}
$k_L a^E$	$k_L a$ at $z = o$ Eq. (97a)	T^{-1}
$k_L a^\alpha$	$k_L a^\alpha$ in range $0.1 \le z \le 1$ Eq. (97c)	T^{-1}
$k_L a_o^\alpha$	$k_L a^\alpha$ at $S = 0.0$	T^{-1}
k_c	coalescence rate constant Eq. (116)	T^{-1}
M_i	relative mole mass of i	—
m_L	coalescence factor Eq. (100)	—
$m = k_L a/(k_L a)_{H_2O}$	coalescence factor Eq. (18)	—
$m^* = (k_L a)_{corr}/(k_L a)_{ref}$	coalescence factor Eq. (19)	—
n	frequency	T^{-1}
O	dissolved oxygen concentration in medium	ML^{-3}
O_B	O in loop	ML^{-3}
O_F	O in tower	ML^{-3}
O_L	O in liquid	ML^{-3}
O^c	critical O in liquid	ML^{-3}
O^*	O at saturation	ML^{-3}
O_F^*	O_F at saturation	ML^{-3}
OTR	oxygen transfer rate	$ML^{-3}T^{-1}$
OUR	oxygen uptake rate	$ML^{-3}T^{-1}$
P	gas pressure	$ML^{-1}T^{-2}$
P_L	integral power input	ML^2T^{-3}
Pr	cell productivity	$ML^{-3}T^{-1}$
Q	tower cross section	L^2
Q_B	loop cross section	L^2
$R(r_x)$	cross correlation function	—
R_c	coalescence rate Eq. (116)	LT^{-1}
R_G	gas constant	$L^2T^{-2}K^{-1}$
$R_x = \dfrac{R_X^*}{X_o + Y_{X/S}S_o}$	dimensionless cell growth rate	—
R_O	oxygen uptake rate	$ML^{-3}T^{-1}$
R_{OB}	oxygen uptake rate in loop	$ML^{-3}T^{-1}$
R_{OF}	oxygen uptake rate in tower	$ML^{-3}T^{-1}$
R_S	substrate uptake rate	$ML^{-3}T^{-1}$
R_{SB}	substrate uptake rate in loop	$ML^{-3}T^{-1}$
R_{SF}	substrate uptake rate in tower	$ML^{-3}T^{-1}$
R_X^*	cell growth rate Eq. (1)	$ML^{-3}T^{-1}$
R_{AB}	cross correlation function Eq. (47)	—
R_E	autocorrelation function Eq. (48)	—
RQ	respiratory quotient	—
r_x	point distance in x direction	L
S	substrate concentration in medium	ML^{-3}
$S^* = \Sigma/\Sigma_R$	foam stability	—
S_B	S in loop	ML^{-3}
S_F	S in tower	ML^{-3}

S_o	S in feed	ML^{-3}
S_D	S at which $k_L a^\alpha = 2k_L a_0^\alpha$	ML^{-3}
T	temperature	K
T_T	turbidity temperature, Eq. (20)	K
t	time	T
U	velocity	LT^{-1}
$\bar{U}$	local mean liquid velocity	LT^{-1}
U_B	bubble rise velocity	LT^{-1}
$U_s = 1 - C_s$	substrate conversion	—
u	velocity measured by anemometer	LT^{-1}
u_x, u_y	turbulence velocity fluctuations	LT^{-1}
u'	turbulence intensity Eq. (45)	LT^{-1}
$u_G = w_{SG}/E_G$	gas velocity	LT^{-1}
$u_B = \dfrac{\dot{V}_R}{Q_B}$	liquid velocity in loop	LT^{-1}
$u_F = w_{SL}/(1 - E_G)$	liquid velocity in tower	LT^{-1}
V	volume of aerated layer in tower	L^3
V_L	volume of bubble free layer in tower	L^3
$\dot{V}_G$	volumetric gas flow rate	L^3T^{-1}
$\dot{V}_L = \dot{V}_R + F$	volumetric liquid flow rate in tower	L^3T^{-1}
$\dot{V}_R$	volumetric recycle rate	L^3T^{-1}
w_{BS}	bubble swarm velocity, Eq. (34)	LT^{-1}
w_{BR}	relative bubble swarm velocity with regard to liquid, Eq. (34a)	LT^{-1}
$w_{SG} = \dfrac{\dot{V}_G}{Q}$	superficial gas velocity in tower	LT^{-1}
$w_{SL} = \dfrac{\dot{V}_L}{Q}$	superficial liquid velocity in tower	LT^{-1}
X	cell mass concentration	ML^{-3}
X_B	X in loop	ML^{-3}
X_F	X in tower	ML^{-3}
X_o	X in feed	ML^{-3}
x	longitudinal coordinate in tower (x = o at aerator)	L
x^*	longitudinal coordinate in loop	L
x_{OG}	mole fraction of O_2 in gas	—
x_{CO_2}	mole fraction of CO_2 in gas	—
x_{OG}^E	x_{OG} at the inlet x = o	—
$Y_{X/O}$	oxygen yield coefficient	—
$Y_{X/S}$	substrate yield coefficient	—
Y	mole fraction	—
$z = x/H$	dimensionless longitudinal coordinate in tower	—
$\gamma = \dfrac{\dot{V}_R}{\dot{V}_L}$	medium recycle ratio	—
η	dynamic viscosity of liquid	$ML^{-1}T^{-1}$
Λ_f	macro scale of turbulence, Eq. (52)	L

λ_f	micro scale of turbulence, Eq. (53)	L
μ_m	maximum specific growth rate	T^{-1}
μ	specific growth rate, Eq. (3)	T^{-1}
ν	kinematic liquid viscosity	L^2T^{-1}
ϱ_L	liquid density	ML^{-3}
ϱ_G	gas density	ML^{-3}
Σ	foaminess, Eq. (27)	T
Σ_R	rest foaminess Eq. (38)	T
σ	surface tension	MT^{-2}
τ	time lag, mean residence time of medium	T
τ_E	dissipation time scale	T
Φ_1	energy dissipation spectrum Eq. (66)	T^{-2}

$$\Psi^\alpha = \frac{k_L a^\alpha}{k_L a^E} \quad \text{coalescence function, Eq. (98)}$$

12 References

1. Schügerl, K.: Chem.-Ing.-Techn. *52*, 951 (1980)
2. Ault, R. G. et al.: J. Inst. Brew. *75*, 261 (1969)
3. Greenshields, R. N., Smith, E. L.: The Chemical Engineer, May 182 (1971)
4. Smith, E. L., Greenshields, R. N.: ibid. Jan. 28 (1974)
5. Royston, M. G.: Process Biochem. *1*, 215 (1966)
6. Klopper, W. J., Roberts, R. H., Royston, M. G.: Proc. Eur. Brew. Conv. 238 (1965)
7. Ault, R. G.: Proc. of tenth Congress, Stockholm, Europ. Brewery Convention, p. 280, Holland: Elsevier 1965
8. Shoore, D. T., Royston, M. G.: Chem. Eng. London, p. 218 May CE 99 (1968)
9. Smith, E. L., James, A., Fidgett, M.: Proc. of Engng. Foundation Conf. on Fluidization, 1978, p. 196, Cambridge: University Press 1978
10. Fidgett, M., Smith, E. L.: Proc. of First Europ. Congress on Biotechn., Part 2 (Poster Papers) 1/92, Interlaken 1978
11. Diesterweg, G., Fuhr, H., Reher, P.: Industrieabwässer Juni p. 7, 1978
12. Bayerturmbiologie (Bayer Tower Biology) Bayer Prospekt E 589-777/68619
13. Schügerl, K., Lücke, J., Oels, U.: In: Bubble Column Bioreactors, Adv. in Biochem. Eng. Vol. 7, p. 1, (Ed.) T. K. Ghose, A. Fiechter, N. Blakebrough, Berlin: Springer 1977
14. Schügerl, K. et al.: Application of tower bioreactors in cell mass production. Adv. in Biochem. Eng. Vol. 8, p. 63, (Ed.) T. K. Ghose, A. Fiechter, N. Blakebrough, Berlin: Springer 1978
15. Lücke, J.: Dissertation, Univ. Hannover, 1976
16. Buchholz, H.: Dissertation, Univ. Hannover 1979
17. Adler, I.: Dissertation, Univ. Hannover 1980
18. Lücke, J., Oels, U., Schügerl, K.: Chem. Ing. Techn. *48*, 573 (1976)
19. Lücke, J., Oels, U., Schügerl, K.: Fifth Int. Fermentation Symp. Berlin 1976, Ed. H. Dellweg, 3 (1976)
20. Lücke, J., Oels, U., Schügerl, K.: Chem. Ing. Techn. *49*, 161 (1977)
21. Schügerl, K.: ibid. *49*, 605 (1977)
22. Schügerl, K., Lücke, J.: Dechema-Monographien Vol. 81 Nr. 1670–1692, 59 (1977)
23. Schügerl, K.: in: „Bioreaktoren" BMFT-Statusseminar „Bioverfahrenstechnik", Stöckheim 1977, Red. H. Keune, R. Scheunemann 155 (1978)
24. Buchholz, H. et al.: Chem. Eng. Sci. *35*, 111 (1980)
25. Schügerl, K.: in: „Bioreaktoren" BMFT-Statusseminar „Bioverfahrenstechnik" Jülich 1979, Red. G. Gartzen, 119 (1979)
26. Lehmann, J. et al.: Dechema Monographien Bd. 81 Nr. 1670–1692, 137 (1977)

218   K. Schügerl

27. Lehmann, J., Hammer, H.: First Europ. Congr. on Biotechn., Part 2 1/73 Interlaken 1978
28. Gerstenberg, H.: Fortschritte der Verfahrenstechnik Bd. 16 Abb. D 1978 (1978)
29. Gerstenberg, H.: Chem. Ing. Techn. *51*, 20 (1979)
30. Mersmann, A.: ibid. *49*, 679 (1977)
31. Deckwer, W.-D.: Fortschritte der Verfahrenstechnik Bd. 15, Abb. D. 303 (1977)
32. Deckwer, W.-D.: Chem. Ing. Techn. *49*, 213 (1977)
33. Yoshida, F., Aktita, K.: A. I. Ch. E. Journal *11*, 9 (1965)
34. Akita, K., Yoshida, F.: Ind. Eng. Chem. Process Des. Dev. *12*, 76 (1973)
35. Akita, K., Yoshida, F.: ibid. *13*, 84 (1974)
36. Buchholz, R., Adler, I., Schügerl, K.: Europ. J. of Appl. Microbiol. Biotechnol. *7*, 135, 241 und 333 (1979)
37. Oels, U., Lücke, J., Schügerl, K.: Chem. Ing. Techn. *49*, 59, (1977)
38. Buchholz, H. et al.: ibid. *50*, 227 (1978)
39. Buchholz, H. et al.: Chem. Eng. Sci. *33*, 1061 (1978)
40. Buchholz, H. et al.: Europ. J. Appl. Microbiol. Biotechnol. *6*, 115 (1978)
41. Katinger, H. W. D., Scheirer, W., Krömer, E.: Chem. Ing. Techn. *50*, 472 (1978)
42. Katinger, H. W. D., Scheirer, W., Krömer, E.: Ger. Chem. Eng. *2*, 31 (1979)
43. Ottmers, D. M., Rase, H. F.: Ind. Engng. Chem. Fundamentals *3*, 106 (1964)
44. Bishoff, K. B., Phillips, J. B.: Ind. Eng. Chem. Proc. Des. Dev. *5*, 416 (1966)
45. Kitai, A., Tone, H., Ozaki, A.: J. Ferment. Technol. *47*, 333 (1969)
46. Kitai, A., Ozaki, A.: ibid. *47*, 527 (1969)
47. Kitai, A., Goto, S., Ozaki, A.: ibid. *47*, 340, 348 and 356 (1969)
48. Kitai, A., Tone, H., Ozaki, A.: Biotech. Bioeng. *11*, 911 (1969)
49. Kitai, A., Yamagata, T.: Process Biochem. *5*, 52 (1970)
50. Prokop, A. et al.: Biotech. Bioeng. *11*, 945 (1969)
51. Zlokarnik, M.: Chem. Ing. Techn. *43*, 329 (1971)
52. Blaß, E., Koch, K. H.: ibid. *44*, 913 (1972)
53. Blaß, E., Cornelius, W.: ibid. *45*, 236 (1973)
54. Cornelius, W., Elstner, F., Onken, U.: Verfahrenstechnik *11*, 304 (1977)
55. Hsu, K. H., Erickson, L. E., Fan, L. T.: Biotech. Bioeng. *17*, 499 (1975)
56. Hsu, K. H., Erickson, L. E., Fan, L. T.: ibid. *19*, 247 (1977)
57. Serieys, M., Goma, G., Durand, G.: ibid. *20*, 1393 (1978)
58. Viesturs, V. E. et al.: ibid. *22*, 799 (1980)
59. Falch, E. A., Gaden, E. L.: ibid. *11*, 927 (1969)
60. Falch, E. A., Gaden, E. L.: ibid. *12*, 465 (1970)
61. Páca, J., Grégr, V.: ibid. *18*, 1075 (1976)
62. Páca, J., Grégr, V.: ibid. *19*, 539 (1977)
63. Páca, J., Grégr, V.: ibid. *21*, 1809 (1979)
64. Páca, J., Grégr, V.: ibid. *21*, 1827 (1979)
65. Páca, J.: Europ. J. Appl. Microbiol. Biotechnol. *9*, 93 (1980)
66. Hackl, A.: Dechema Monographien Vol. 73, Nr. 1410–1431 37 (1973)
67. Strébácek, Z., Sáchová, M.: Chem. Ing. Techn. *46*, 397 (1974)
68. Brauer, H., Schmidt-Traub, H., Thiele, H.: ibid. *46*, 699 (1974)
69. Wen-Jei Yang: Letters in Heat and Mass Transfer, Pergamon Press *3*, 433 (1976)
70. Brauer, H.: Dechema-Monographien Vol. 81 Nr. 1670–1692 87 (Biotechn. Symp. Tutzing) 1977
71. Meister, D. et al.: Chem. Eng. Sci. *34*, 1367 (1979)
72. Brauer, H., Thiele, H.: Chem. Anlagen Verfahren Nr. 4, 32 (1975)
73. Brauer, H., Sucker, D.: Chem. Ing. Techn. *50*, 876 (1978)
74. Al. Taweel, A. M., Landau, J., Picot, J. J. C.: 85th A. I. Ch. E. Meet. Jan. 1978
75. Miyanami, K. et al.: Chem. Eng. Sci. *33*, 601 (1978)
76. McLean, G. T. et al.: Biotech. Bioeng. *19*, 493 (1977)
77. Gutierrez, J. R., Erickson, L. E.: ibid. *20*, 487 (1978)
78. Wang, K. B., Fan, L. T.: Chem. Eng. Sci. *33*, 945 (1978)
79. Leistner, G., Zibinski, E.: Industrieabwässer, June 13–20 (1978)
80. Voigt, J., Schügerl, K.: Chem. Ing. Techn. *50*, 721 (1978)

81. Schügerl, K.: Verwendung von Blasensäulen als Bioreaktoren, Rothenburger Symp. Fermentationstechnik Juli 1978, Leit. R. L. Lafferty, Braun Melsungen, 11 (1978)
82. Voigt, J., Schügerl, K.: Chem. Eng. Sci. *34*, 1221 (1979)
83. Schügerl, K.: Journées d'études gaz-liquide transfer de matérie agitation et mélange, INSA Toulouse 12–13. Septembre 1979. Resp. M. M. Roques, M. Roustan IV. 3, 1979
84. Voigt, J., Hecht, V., Schügerl, K.: Chem. Eng. Sci. *35*, 1317 (1980)
85. Hecht, V., Voigt, J., Schügerl, K.: ibid. *35*, 1325 (1980)
86. Schügerl, K.: Verfahrenstechnik *14*, 727 (1980)
87. Voigt, J.: Dissertation, Univ. Hannover 1980
88. Voigt, J., Schügerl, K.: Europ. J. of Appl. Microbiol. Biotechn. *11*, 97 (1981)
89. Voigt, J., Schügerl, K.: Chem. Ing. Techn. *52*, 995 (1980)
90. Voigt, J., Schügerl, K.: Proc. of Int. Fermentation Symp., London, Ont., Can. 1980
91. Herbrechtsmeier, P., Steiner, R.: Chem. Ing. Techn. *50*, 944 (1978)
92. Albrecht, W. J., Schulze-Pillot, G.: ibid. *50*, 947 (1978)
93. Stein, W.: ibid. *40*, 829 (1968)
94. Lehnert, J., Niewerth, E.: Verfahrenstechnik *3*, 382 (1969)
95. Blenke, H., Bohner, K., Hirner, W.: ibid. *3*, 444 (1969)
96. Blenke, H., Bohner, K., Hirner, W.: Chem. Ing. Techn. *42*, 479 (1970)
97. Blenke, H., Bohner, K., Pfeiffer, W.: ibid. *43*, 10 (1971)
98. Bohner, K., Blenke, H.: Verfahrenstechnik *6*, 50 (1972)
99. Lehnert, J.: ibid. *6*, 58 (1972)
100. Hirner, W., Blenke, H.: Chem. Ing. Techn. *46*, 352 (1974)
101. Blenke, H., Hirner, W.: VDI-Berichte *218*, 549 (1974)
102. Blenke, H.: Schlaufenreaktoren. Bioreaktoren. BMFT Statusseminar „Bioverfahrenstechnik", Stöckheim 1977. Red. H. Keune, R. Scheunemann, 5 (1977)
103. Hirner, W.: Dissertation, Univ. Stuttgart 1974
104. Hirner, W., Blenke, H.: Verfahrenstechnik *11*, 297 (1977)
105. Marquart, R.: ibid. *13*, 527 (1979)
106. Seipenbusch, R., Birckenstädt, J. W., Schindler, F.: 1. Symp. Mikrobielle Proteingewinnung, 1975 GBF-DGHM, Verlag Chemie, Weinheim, 59 (1975)
107. Seipenbusch, R. et al.: Fifth Int. Fermentation Symp., Berlin 1976, Ed. H. W. Dellweg, 4.09 (1976)
108. Knecht, R. et al.: Process Biochemistry, May, 11 (1977)
109. Birckenstädt, J. W., Faust, U., Sambeth, W.: ibid. Nov., 7 (1977)
110. Faust, U.: Proc. XII. Int. Congr. of Microbiol. München 1978, Dechema Monographien Bd. 83 Nr. 1704–1723, 125 (1979)
111. Sittig, W., Heine, H.: Chem. Ing. Techn. *49*, 595 (1977)
112. Blenke, H.: Loop Reactors, in: Adv. Biochem. Eng. Vol. 13, p. 121, (Ed.) T. K. Ghose, A. Fiechter, N. Blakebrough, Berlin: Springer 1979
113. Seipenbusch, R., Blenke, H.: The Loop Reactor for Cultivating Yeast on n-Paraffin Substrate, in: Adv. Biochem. Eng. Vol. 15, p. 1, (Ed.) A. Fiechter, Berlin: Springer 1980
114. Blenke, H., Reule, W., Schum, W.: in: Bioreaktoren, 2. BMFT-Statusseminar „Bioverfahrenstechnik", Jülich 1979, Red. J. Gartren, 273 (1979)
115. Le François, L.: Chim. Ind. *8*, 1038 (1969)
116. Hatch, R. T.: Dissertation MIT, Cambridge Mass. 1973
117. Hatch, R. T.: in: „Single Cell Protein" II. Ed. S. R. Tannenbaum, D. I. C. Wang, M. I. T. Press, 46 (1975)
118. Wang, D. I. C., Hatch, R. T., Cuevas, C.: Engineering Aspects of SCP Production from Hydrocarbon Substrates: The Airlift Fermenter Proc. 8th World Petroleum Congr., Moskau 1971, 149 (1971)
119. Laine, B., Vernet, C., Evans, G.: Progress Recents dans la Production de Proteins a Partir Petrole. Proc. 7th World Petroleum Congr., Mexico City 1967
120. Laine, B., du Chaffaut, J.: in: „Single Cell Protein" II. Ed. S. R. Tannenbaum, D. I. C. Wang, M. I. T. Press, 424 (1975)
121. Cooper, P. G., Silver, R. S., Boyle, J. B.: in: „Single Cell Protein" II. Ed. S. R. Tannenbaum, D. I. C. Wang, M. I. T. Press, 454 (1975)

122. Wise, D. L., Wang, D. I. C., Racicot, H. A.: Chem. Engng. Progr. Sympos. Ser. Nr. 107 Vol. 67, 554 (1971)
123. Le François, L., Revus, B.: Dechema Monographien Bd. 70 No. 1327—1350, 97 (1972)
124. Champagnat, A.: Chem. Engng. Vo. 26 Nov., 62 (1973)
125. Laine, B.: Can. J. Chem. Engng. 50, 154 (1972)
126. Rauschenberger, J., Nagy, Z., Kovacs, A.: Dechema-Monographien Bd. 80 No. 1616–1638, 129 (1976)
127. Euzen, J. P., Trambouze, P., van Landeghem, H.: ISCRE 5, Houston 153-(1978)
128. Laine, J., Knoppamäkl, R.: Ind. Eng. Chem. Process Des. Dev. 18, 3, 501 (1979)
129. Gasner, L. L.: Biotech. Bioeng. 16, 1179 (1974)
130. Bolton, D. H., Hines, D. A., Bouchard, J. P.: The application of the ICI deep shaft process to industrial effluents, p. 344. 31st Annual Purdue Ind. Waste Conf., Lafayette, Indiana 1976
131. Ho, C. S., Erickson, L. E., Fan, L. T.: Biotech. Bioeng. 19, 1503 (1977)
132. Pilepp, E., Scheffler, U., Schmidt-Mende, P.: „Bioreaktoren", BMFT-Statusseminar „Bioverfahrenstechnik", Stöckheim 1977, (Red.) H. Keune, R. Scheunemann, 45 (1977)
133. Hines, D. A.: Dechema-Monographien, Biotechn. Proc. 1st Europ. Congr. on Biotechn. Interlaken 1978. Survey Lectures Vol. 82 No. 1693–1703, 55 (1978)
134. Müller, G. et al.: Chemie Technik 7 No. 6, 257 (1978)
135. Plants for manufacturing bioprotein, Uhde Prospect Lo II 519 500 79
136. Scheffler, U., Buchel, M., Pilepp, E.: in: „Bioreaktoren", 2. BMFT-Statusseminar „Bioverfahrenstechnik" Jülich 1979, Red. J. Gartzen, p. 233 (1979)
137. Faust, U., Sittig, W.: Methanol as Carbon Source for Biomass Production in a Loop Reactor, Adv. in Biochem. Engng. Vol. 17, p. 63. (Ed.) A. Fiechter, Berlin: Springer 1980
138. Margaritis, A., Kennedy, K., Zajic, J. E.: Developm. in Ind. Microbiology 21, 285 (1980)
139. Hines, D. A. et al.: I. Chem. E. Symp. Series No. 41 D1 (1975)
140. Full scale single cell protein plant. Process Biochem. Jan./Febr., 30 (1977)
141. Orazem, M. E., Erickson, L. E.: Biotech. Bioeng. 21, 69 (1979)
142. Orazem, M. E., Fan, L. T., Erickson, L. E.: ibid. 21, 1579 (1979)
143. Littlehailes, J. D.: 1. Symp. Mikrobielle Proteingewinnung GBF, DGHM 1975, Verlag Chemie, 43 (1975)
144. Cow, J. S. et al.: „Single Cell Protein" II (Ed.) S. R. Tannenbaum, D. I. C. Wang, M. I. T. Press, 370 (1975)
145. Kanazawa, M.: „Single Cell Protein" II. (Ed.) S. R. Tannenbaum, D. I. C. Wang, M. I. T. Press, 438 (1975)
146. Lin, C. H. et al.: Biotech. Bioeng. 18, 1557 (1976)
147. Weiland, R., Onken, U.: in: „Bioreaktoren", BMFG-Statusseminar „Bioverfahrenstechnik", Stöckheim 1977, (Red.) H. Keune, R. Scheunemann, 65 (1977)
148. Weiland, P.: Dissertation, Univ. Dortmund 1978
149. Weiland, P., Brentrup, L., Onken, U.: „Bioreaktoren", 2. BMFT-Statusseminar „Bioverfahrenstechnik" Jülich 1979 Red. J. Gartzen, 249 (1979)
150. Weiland, P., Onken, U.: Chem. Ing. Techn. 52, 264 (1980)
151. Onken, U., Weiland, P.: Europ. J. Appl. Microbiol. Biotechnol. 10, 31 (1980)
152. Bergeys Manual of Determinative Bacteriology 8th Ed., p. 293. The Williams & Wilkins Company, Baltimore 1974
153. Buchholz, H. et al.: Chem. Ing. Techn. 52, 836 (1980)
154. Buchholz, H. et al.: Europ. J. Appl. Microbiol. Biotechnol. 11, 89 (1981)
155. Buchholz, H. et al.: J. of Chem. Technol. and Biotechn. 31, 435 (1981)
156. Müllner, J.: Waagner-Biró-Submerging Channel Reactor for Waste Water Treatment. Presented in the 5th Working Session of Chem. Engng., Graz 1977
157. Müllner, J.: Waagner-Biró GmbH. O. E. Pat. 319864 (1973)
158. Mitchell, R. C., Lev, A. D.: Chem. Engng. Progress Symp. Series No. 107 Vol. 67, 558 (1970)
159. Katinger, H. W. D.: 3. Symp. Techn. Mikrobiologie Berlin, 1973, 95 (1973)
160. Leuteritz, G. M., Reimann, P., Vergeres, P.: Hydrocarbon Processing, June 99 (1976)
161. van de Sande, E., Smith, J. M.: Chem. Ing. Techn. 44, 1177 (1972)
162. van de Sande, E., Smith, J. M.: Chem. Eng. Sci. 28, 1161 (1973)
163. van de Sande, E., Smith, J. M.: ibid. 31, 219 (1976)

164. Suicu, G. D., Smigelschi, O.: ibid. *31*, 1217 (1976)
165. Burgess, J. M., Molloy, N. A.: ibid. *27*, 442 (1972)
166. Burgess, J. M., Molloy, N. A.: ibid. *28*, 183 (1973)
167. van de Donk, J., Smith, J. M.: Air entrainment by plunging water jets — Large scale data. 6th CHISA Congr. A. 4.3 (1978)
168. Langhans, G., Liepe, F., Richter, K.: Chem. Techn. *26*, 519 (1974)
169. Schönherr, W., Jagusch, L.: ibid. *28*, 687 (1976)
170. Langhans, G. et al.: Chem. Techn. *29*, 662 (1977)
171. Liepe, G. et al.: First Europ. Congr. on Biotechn., Part 1, 78, Interlaken 1978
172. Schreier, K.: Chem. Ztg. *99*, 328 (1975)
173. Schreier, K.: Fifth Int. Fermentation Symp., Berlin 1976, 1.03 (1976)
174. Schreier, K.: First Europ. Congr. on Biotechn., Part 2, 1/67, Interlaken 1978
175. Lafferty, R. M. et al.: Chem. Ing. Techn. *50*, 401 (1978)
176. Lafferty, R. M. et al.: Plunging jet loop reactors, paper presented on the Dechema Jahrestagung 1977
177. Moebus, O., Teuber, M.: Kieler Milchwirtschaftliche Forschungsberichte *31*, 297 (1979)
178. Adler, I., Lippert, J., Schügerl, K.: Rothenburger Symp. Karlshafen 1980
179. Kuraishi, M. et al.: Microbiology applied to Biotechn., Proc. XII. Int. Congr. on Microbiol. München Sept. 3–8, 1978. Dechema Monographien Bd. 83 Nr. 1704–1723, 111 (1979)
180. Goma, G.: Agitation-Aeration de Cultures de Microorganisms. Journées d'études gas-liquide transfert de matière agitation et mélange, Toulouse 13th Sept. 1979
181. Luttmann, R.: Dissertation, Univ. Hannover 1980
182. Atkinson, B., Daoud, I. S.: Microbial Flocs and Flocculation in Fermentation Process Engineering, Adv. in Biochem. Engng. (Eds.) T. K. Ghose, A. Fiechter, N. Blakebrough, Vol. 4, p. 41. Heidelberg: Springer 1976
183. Bumbullis, W., Kalischewski, K., Schügerl, K.: Europ. J. Appl. Microbiol. Biotechnol. *7*, 147 (1979)
184. Ziminski, S. A., Caron, M. M., Blackmore, R. B.: Ind. Eng. Chem. Fundam. *6*, 233 (1967)
185. Hassan, I. T. M., Robinson, C. W.: Chem. Eng. Sci. *35*, 1277 (1980)
186. Zieminski, S. A., Whittemore, R. C.: ibid. *26*, 509 (1971)
187. Lessard, R. R., Zieminski, S. A.: Ind. Eng. Chem. Fundam. *10*, 260 (1971)
188. Zlokarnik, M.: Sorption Characteristics for Gas-Liquid Contacting in Mixing Vessels. Adv. in Biochem. Engng. (Eds.) T. K. Ghose, A. Fiechter, N. Blakebrough, Vol. 8, p. 133. Berlin: Springer 1978
189. Bernal, J. D., Fowler, R. H.: J. Chem. Phys. *1*, 515 (1933)
190. Kavanau, L. J.: Water and Solute-Water Interactions, Holden-Day 1964
191. Hofmeister, F.: Arch. Exp. Pathol. Pharmakol. *24*, 247 (1888)
192. Luck, W.: Fortschr. Chem. Forschung, Springer Verlag Berlin, *4*, 653 (1964)
193. Luck, W.: Top. Curr. Chem., Springer Verlag Berlin, *64*, 113 (1976)
194. Raymond, D. R., Zieminski, S. A.: A. I. Ch. E. Journal *17*, 57 (1971)
195. Sagert, N. H. et al.: Proc. of a Symp. on Foams at the Brunel Univ. Sept. 8–10, 1975. Ed. R. J. Ahers, Acad. Press, 147, (1976)
196. Bumbullis, W., Schügerl, K.: Europ. J. Appl. Microbiol. Biotechn. *8*, 17 (1979)
197. Elstner, F., Onken, U.: Chem. Ing. Techn. *52*, 597 (1980)
198. Danckwerts, P. J.: Gas-liquid reactions, McGraw Hill Co. 1970
199. Schumpe, A., Deckwer, W.-D.: Biotech. Bioeng. *21*, 1079 (1979)
200. Schumpe, A., Deckwer, W.-D.: Oxygen solution in synthetic fermentation media. First Europ. Congr. on Biotechn., M/z, 154, Interlaken 1978
201. Quicker, G.: Diplomarbeit, Univ. of Hanover 1980
202. Popovic, M., Niebeschütz, H., Reuß, M.: Europ. J. Appl. Microbiol. Biotechnol. *8*, 1 (1979)
203. Quicker, G. et al.: Biotech. Bioeng. *23*, 635 (1981)
204. Heyduk, H., Chang, S. C.: Chem. Eng. Sci. *26*, 635 (1971)
205. Lohse, M. et al.: A. I. Ch. E. Journal *27*, 626 (1981)
    Lohse, M. et al.: Solubilities and Diffusivities of CO_2 in Dilute Polymeric Solutions. Poster. Proc. of Phase Equilibria and Fluid Properties in the Chemical Industry. 2nd Int. Conf. Berlin (West), 17–21 March 1980
206. Young, M. E., Carroad, P. A., Bell, R. L.: Biotech. Bioeng. *22*, 947 (1980)

207. Reuß, M. et al.: Europ. J. Appl. Microbiol. Biotechn. *8*, 167 (1979)
208. Davies, J. T., Rideal, E. K.: Interfacial Phenomenoa, New York: Acad. Press 1963
209. König, B., Kalischewski, K., Schügerl, K.: Europ. J. Appl. Microbiol. Biotechn. *7*, 251 (1979)
210. Buchholz, H., Kalischewski, K., Schügerl, K.: ibid. *7*, 321 (1979)
211. Lucassen, J., van den Tempel, M.: Chem. Eng. Sci. *27*, 1283 (1972)
212. Lucassen, J., van den Tempel, M.: J. Colloid and Interface Science *41*, 491 (1972)
213. Bumbullis, W.: Dissertation, Univ. Hannover 1980
214. Bumbullis, W., Schügerl, K.: Part VII Europ. J. Appl. Microbiol. Biotechn. *11*, 110 (1981)
215. Zlokarnik, M.: Influence of various material and process related parameters on bubble coalescence in gas/liquid contacting. First Europ. Congr. on Biotechn. Part I. 1/13, Interlaken 1978
216. Keitel, G.: Dissertation, Univ. Dortmund 1978
217. Adler, I. et al.: Europ. J. Appl. Microbiol. Biotechn. *9*, 249 (1980)
218. Adler, I. et al.: Europ. J. Appl. Microbiol. Biotechn. *10*, 171 (1980)
219. Cumper, C. W. N., Alexander, A. E.: Trans. Farad. Soc. *46*, 235 (1950)
220. Kalischewski, K., Bumbullis, W., Schügerl, K.: Europ. J. Appl. Microbiol. Biotechn. *7*, 21 (1979)
221. Bikermann, J.: Trans. Faraday Soc. *34*, 634 (1938)
222. Bumbullis, W., Schügerl, K.: Part VI. Europ. J. Appl. Microbiol. Biotechn. *11*, 106 (1981)
223. Burgess, J. M., Calderbank, P. H.: Chem. Eng. Sci. *30*, 743, 1107, 1511 (1975)
224. Raper, J. A. et al.: Chem. Eng. Sci. *33*, 1405 (1978)
225. Calderbank, P. H.: ibid. *33*, 1407 (1978)
226. Buchholz, R.; Schügerl, K.: Europ. J. Appl. Microbiol. Biotechn. *6*, 301, 315 (1979)
227. Buchholz, R., Zakrzewski, W., Schügerl, K.: Chem. Ing. Techn. *51*, 568 (1979)
228. Brentrup, L., Weiland, P., Onken, U.: ibid. *52*, 72 (1980)
229. Brentrup, L., Weiland, P., Onken, U.: German Chem. Engng. *3*, 296 (1980)
230. Sridhar, T., Potter, O. E.: Chem. Eng. Sci. *35*, 683 (1980)
231. Calderbank, P. H., Pereira, J.: ibid. *32*, 1427 (1977)
232. Calderbank, P. H.: Trans. Inst. Chem. Engrs. *36*, 443 (1958)
233. Hofer, H., Mersmann, A.: Chem. Ing. Techn. *52*, 362 (1980)
234. Shridar, T., Potter, O. E.: Chem. Eng. Sci. *33*, 1347 (1978)
235. Schumpe, A., Deckwer, W.-D.: Chem. Ing. Techn. *52*, 468 (1980)
236. Van Dijken, J. P. et al.: Arch. Microbiol. *105*, 261 (1975)
237. Van Dijken, J. P. et al.: Arch. Microbiol. *102*, 41 (1975)
238. Van Dijken, J. P., Otto, R., Harder, W.: ibid. *111*, 137 (1976)
239. Van Dijken, J. P.: Dissertation, Univ. Groningen 1976
240. Eggeling, L., Sahm, H.: Eur. J. App. Microbiol. Biotechn. *5*, 197 (1978)
241. Eggeling, L., Sahm, H.: Arch. Microbiol. *127*, 119 (1980)
242. Egli, T.: Dissertation E. T. H. Zürich Nr. 6538, 1980
243. Rubin, E., Gaden, E. L. Jr.: Foam Separation in: „New Chemical Engineering Separation Techniques", (Ed.) H. M. Schoen, New York: Interscience Publisher 319, 1962
244. Lehmlich, R. (Ed.): Adsorptive Bubble Separation Techniques, New York: Acad. Press 1972
245. Somasundaran, P.: Foam Separation Methods, in Separation and Purification Methods *1*, 117 (1972)
246. Grieves, R. B.: The Chem. Engng. J. *38*, 183 (1974)
247. Urrutia Desmaison, G., Schügerl, K.: Chem. Ing. Techn. *52*, 855 (1980)
248. Windish, S.: Monatsschr. f. Brauerei *22*, 69 (1969)
249. Reif, F. et al.: in: Die Hefen in der Wissenschaft, Vol. I, p. 527. Hans Carl Verlag 1960
250. Mill, P. J.: J. Gen. Microbiol. *35*, 53, 61 (1964)
251. Lyons, T. P., Hough, J. S.: J. Inst. Brew *76*, 564 (1970); *77*, 300 (1971)
252. Baker, D. A., Kirsop, B. H.: ibid. *78*, 454 (1972)
253. Kalynzhnyi, M. Ya.: Microbiology *31*, 586 (1962)
254. Patel, G. B., Ingledew, W. M.: J. Inst. Brew. *81*, 123 (1975)
255. Patel, G. B., Ingledew, W. M.: Can. J. Microbiol. *21*, 1608, 1614 (1975)
256. Kuhlmann, W. et al.: unpublished

257. Kirstensen, H. S.: Hot wire measurements in turbulent flows DISA-Information Department DK-2740 Skovlunde 1974
258. Zakrzewski, W. et al.: Part IV. Europ. J. of Appl. Microbiol. Biotechn. *12*, 69 (1981)
259. Zakrzewski, W. et al.: Chem. Ing. Techn. *53*, 135 (1981)
260. Zakrzewski, W. et al.: Part VI. Europ. J. of Appl. Microbiol. Biotechn. *12*, 150 (1981)
261. Lippert, J. et al.: Chem. Ing. Techn. *53*, 967 (1981)
262. Hinze, J. O.: Turbulence, McGraw-Hill Co. 1975
263. Bendat, J. S., Piersol, A. G.: Random Data, Wiley Interscience 1971
264. Taylor, G. I.: Proc. Roy. Soc., London *A 151*, 421 (1935)
265. Kolmogoroff, I. N.: Compt. rend. acad. sci. RSS *30*, 301; *32*, 16 (1941)
266. v. Weizsäcker, C. F.: Z. f. Phys. *124*, 614 (1948)
267. Heisenberg, W.: ibid. *124*, 628 (1948)
268. v. Karman, Th., Howard, L.: Proc. Roy. Soc. London *A 164*, 192 (1938)
269. Luttmann, R. et al.: Part I. Computers and Chem. Eng. 6 (1982)
270. Luttmann, R. et al.: Part II. Computers and Chem. Eng. 6 (1982)
271. Luttmann, R. et al.: Biotech. Bioeng. 24 (1982)
272. Scheiding, W.: Diplomarbeit, Univ. Hannover 1980
273. Zakrzewski, W.: Dissertation, Univ. Hannover 1980
274. Zakrzewski, W. et al.: Part V. Europ. J. of Appl. Microbiol. Biotechn. *12*, 143 (1981)
275. Alvarez-Cuenca, M., Baker, C. G. J., Bergougnou, M. A.: Chem. Eng. Sci. *35*, 1121 (1980)
276. Deckwer, W.-D., Hallensleben, J., Popovic, M.: Can. J. of Chem. Engng. *58*, 190 (1980)
277. Burckhart, R., Deckwer, W.-D.: Verfahrenstechn. *10*, 429 (1976)
278. Deckwer, W.-D., Burckhart, R., Zoll, G.: Chem. Eng. Sci. *29*, 2177 (1974)
279. Deckwer, W.-D., Adler, I., Zaidi, A.: Can. J. Chem. Eng. *56*, 43 (1978)
280. Hallensleben, J. et al.: Chem. Ing. Techn. *49*, 663 (1977)
281. Brauer, H.: ibid. *51*, 934 (1979)
282. Buchholz, H. et al.: Europ. J. Appl. Microbiol. Biotechn. *12*, 63 (1981)
283. Oels, U.: Dissertation, Univ. Hannover 1975
284. Adler, I. et al.: Part III. Europe J. Appl. Microbiol. Biotechn. *12*, 212 (1981)
285. Chen, M. S. K.: A. I. Ch. E. Journal *18*, 849 (1972)
286. Pasquali, G., Magelli, F.: Chem. Eng. Journal *9*, 83 (1975)
287. Shioya, S., Dang, N. D. P., Dunn, U. J.: Chem. Eng. Sci. *33*, 1025 (1978)
288. Merchuk, J. C., Stein, Y., Mateles, R. I.: Biotech. Bioeng. *22*, 1189 (1980)
289. Shioya, S. et al.: 6th Int. Fermentation Symp., London, Ont. Can. F7.3.2(L), 1980
290. Whalley, P. B., Davidson, J. F.: Proc. of the Symp. on Multiphase Flow Systems, Glasgow, Scotland, Ser. No. 38, J5 (1974)
291. Joshi, J. B., Sharma, M. M.: Can. J. Chem. Engng. *57*, 375 (1979)
292. Diekmann, J.: Diplomarbeit, Univ. Hannover 1981
293. Badura, R. et al.: Chem. Ing. Techn. *46*, 399 (1974)
294. Mangarz, K. H., Pilhofer, Th.: Verfahrenstechnik *14*, 40 (1980)
295. Adler, I., Deckwer, W.-D., Schügerl, K.: Chem. Eng. Sci. I+II *37*, 271, 417 (1982)
296. Chen, G. K. C., Fan, L. T., Erickson, L. T.: Can. J. Chem. Eng. *50*, 157 (1972)
297. Todt, J. et al.: Chem. Eng. Sci. *32*, 369 (1977)
298. Pirt, S. J.: Principles of Microbe and Cell Cultivation, Blackwell Scientific Publications 1975
299. Bishoff, K. B.: Canad. J. Chem. Engng. *44*, 281 (1966)
300. Kolmogoroff, A. N.: Compt. rend. acad. sci. USSR *30*, 301 (1941)
301. Calderbank, P. H.: in: Biochem. and Biological Engng. Science Vol. 1, p. 101, 1967. (Ed.) N. Blakebrough, Acad. Press
302. Zlokarnik, M.: Korrespondenz Abwasser *27*, 194 (1980)
303. Adler, I. et al.: 2. Symp. Microbielle Proteingewinnung. GBF Braunschweig-Stöckheim: Verlag Chemie 1980 Ed. P. Präve, K. Schügerl, H. Zucker, p. 15
304. Reith, T.: Doctoral Thesis, Technische Hogeschool Delft 1968
305. Adler, I., Schügerl, K.: Biotech. Bioeng. (in press)
306. Schlegel, H. G.: Allgem. Mikrobiologie, Stuttgart: Thieme Verlag 1976
307. Bergeter, F.: Wachstum von Mikroorganismen, Jena: Fischer Verlag 1972
308. Thoma, M., Luttmann, R.: in: „Bioreaktoren" 2. BMFT-Statusseminar „Bioverfahrenstechnik", Jülich 1979, 199 (1979)

309. Munack, A.: Dissertation, Univ. Hannover 1980
310. Munack, A., Thoma, M.: in: „Summer School on Modelling of Dynamical Systems based on Experimental Data with Chemical Engineering Applications. (Ed.) A. Pethö. Bad Honnef, Aug. 17–29, 1980
311. Schlingmann, H.: Dissertation, Univ. Hannover 1980
312. Schlingmann, H. et al.: Computers and Chem. Engng. *3*, 53 (1979)
313. Luttmann, R. et al.: Mathematical treatment of SCP processes in tower loop reactors (to be submitted to Adv. Biochem. Eng.)
314. Diderichsen, B., J. Bacteriol. 858 (1980)
315. Luttmann, R. et al.: I+II Europ. J. Appl. Microbiol. Biotechn. *13*, 90, 145 (1981)
316. Luttmann, R. et al.: Chem. Eng. Sci. 37 (1982)

Author Index Vol. 1–22

H. Brauer, Y. B. G. Varma

Air Pollution Control Equipment

1981. 285 figures, 53 tables. VII, 388 pages
ISBN 3-540-10463-1

Contents: Introduction to the Problems of Environmental Protection. – Integration of Technical Measures Taken for Environmental Protection. – Survey on Technical Processes and Equipment for Air Pollution Control and Some Fundamentals. – Design and Operation of Cyclones. – Design and Operation of Wet Dust Scrubbers. – Design and Operation of Fabric Filters. – Design and Operation of Electrical Precipitators. – Design and Operation of Mist Separators. – Design and Operation of Absorption Equipment. – Design and Operation of Adsorption Equipment. – Design and Operation of Equipment for Biological Waste Gas Treatment. – Design and Operation of Equipment for Chemical Waste Gas Treatment. – Subject Index.

Air Pollution Control Equipment is a critical compilation of available knowledge in the design and operation of equipment for abating harmful emissions. Air pollution control technology, though of relatively recent origin, has already achieved a high degree of efficacy due to the intense research and development invested in it in past decades.

This book is addressed primarily to chemical, environmental, and machanical engineers who are engaged in the design and operation of equipment for air pollution control. But it will certainly be helpful to chemists and physicists who are confronted with the solutions of environmental problems. Furthermore it is intended as a textbook engineering courses on environmental protection.

Springer-Verlag
Berlin
Heidelberg
New York

E. Lück

Antimicrobial Food Additives

Characteristics, Uses, Effects

Translated from the German by G. F. Edwards

1980. 1 figure, 37 tables. XVIII, 280 pages
ISBN 3-540-10056-3

Contents: General considerations: Aim and development of food preservation. Analytical detection of preservatives. Health aspects. The legal situation relating to food. Antimicrobial action of preservatives. – The individual preservatives: Sodium chloride. Silver. Boric acid. Carbon dioxide. Nitrogen. Nitrates. Nitrites. Ozone. Hydrogen peroxide. Sulfur dioxide. Chlorine. Ethyl alcohol. Ethylene oxide. Sucrose. Hexamethylenetetramine. Formic acid. Acetic acid. Propionic acid. Sorbic acid. Dehydroacetic acid. Dicarbonic acid esters. Benzoic acid. Salicylic acid. Esters of p-hydroxybenzoic acid. o-Phenylphenol. Diphenyl. Smoke. Furyl furamide. Thiabendazole. Nisin. Pimaricin. Other preservatives. Packagings and coatings. – Subject index.

The chemical preservation of food plays an essential role both in food science research and food industry processes. Despite its importance, however, information on the subject remained dispersed in a number of journals and handbooks for a long time. The foremost reference works on food microbiology and technology treated preservation by chemical means only peripherally.

The original German edition of this book was the first comprehensive monograph to appear on the subject. Highly acclaimed by scientists throughout Europe, it is now available to English-speaking audiences in expanded, updated translation. General considerations relevant to all types of preservatives are treated in the opening section. The second part contains descriptions of individual preservatives themselves according to the system employed in organic and inorganic chemistry. Separate chapters are devoted to those substances of major practical importance, including disinfectants and gasses used for degermination.

This book's detailed explanations of the principles and interrelationships relevant to the practical use of preservatives in the food sector, as well as the comprehensive literature reference it provides, will make it an invaluable source of information for researchers, government authorities, physicians, nutrition scientists and interested laymen.

Springer-Verlag
Berlin
Heidelberg
New York